Lecture Notes in Computer Science 16464

Founding Editors

Gerhard Goos
Juris Hartmanis

Editorial Board Members

Elisa Bertino, *Purdue University, West Lafayette, IN, USA*
Wen Gao, *Peking University, Beijing, China*
Bernhard Steffen, *TU Dortmund University, Dortmund, Germany*
Moti Yung, *Columbia University, New York, NY, USA*

The series Lecture Notes in Computer Science (LNCS), including its subseries Lecture Notes in Artificial Intelligence (LNAI) and Lecture Notes in Bioinformatics (LNBI), has established itself as a medium for the publication of new developments in computer science and information technology research, teaching, and education.

LNCS enjoys close cooperation with the computer science R & D community, the series counts many renowned academics among its volume editors and paper authors, and collaborates with prestigious societies. Its mission is to serve this international community by providing an invaluable service, mainly focused on the publication of conference and workshop proceedings and postproceedings. LNCS commenced publication in 1973.

Mauricio Reyes · Pedro Henriques Abreu ·
Jaime Cardoso

Editors

Interpretability of Machine Intelligence in Medical Image Computing

8th International Workshop, iMIMIC 2025
Held in Conjunction with MICCAI 2025
Daejeon, South Korea, September 27, 2025
Proceedings

 Springer

Editors
Mauricio Reyes
University of Bern
Bern, Switzerland

Pedro Henriques Abreu
CISUC
Coimbra, Portugal

Jaime Cardoso
INESC TEC
Porto, Portugal

ISSN 0302-9743 ISSN 1611-3349 (electronic)
Lecture Notes in Computer Science
ISBN 978-3-032-17610-3 ISBN 978-3-032-17611-0 (eBook)
https://doi.org/10.1007/978-3-032-17611-0

© The Editor(s) (if applicable) and The Author(s), under exclusive license
to Springer Nature Switzerland AG 2026

This work is subject to copyright. All rights are solely and exclusively licensed by the Publisher, whether the whole or part of the material is concerned, specifically the rights of translation, reprinting, reuse of illustrations, recitation, broadcasting, reproduction on microfilms or in any other physical way, and transmission or information storage and retrieval, electronic adaptation, computer software, or by similar or dissimilar methodology now known or hereafter developed.
The use of general descriptive names, registered names, trademarks, service marks, etc. in this publication does not imply, even in the absence of a specific statement, that such names are exempt from the relevant protective laws and regulations and therefore free for general use.
The publisher, the authors and the editors are safe to assume that the advice and information in this book are believed to be true and accurate at the date of publication. Neither the publisher nor the authors or the editors give a warranty, expressed or implied, with respect to the material contained herein or for any errors or omissions that may have been made. The publisher remains neutral with regard to jurisdictional claims in published maps and institutional affiliations.

This Springer imprint is published by the registered company Springer Nature Switzerland AG
The registered company address is: Gewerbestrasse 11, 6330 Cham, Switzerland

If disposing of this product, please recycle the paper.

Preface

This book constitutes the refereed proceedings of the 8th International Workshop on Interpretability of Machine Intelligence in Medical Image Computing, iMIMIC 2025, held on September 27th, 2025, in conjunction with the 28th International Conference on Medical Imaging and Computer-Assisted Intervention, MICCAI 2025. iMIMIC is a single-track, half-day workshop consisting of high-quality, previously unpublished papers, presented either orally or as a poster, and is intended to act as a forum for research groups, engineers, and practitioners to present recent algorithmic developments, new results, and promising future directions in interpretability of machine intelligence in medical image computing. Machine learning systems are achieving remarkable performances at the cost of increased complexity. Hence, they become less interpretable, which may cause distrust, potentially limiting clinical acceptance. As these systems are pervasively being introduced to critical domains, such as medical image computing and computer-assisted intervention, it becomes imperative to develop methodologies allowing insight into their decision making. Such methodologies would help physicians to decide whether they should follow and trust automated decisions. Additionally, interpretable machine learning methods could facilitate defining the legal framework of their clinical deployment. Ultimately, interpretability is closely related to AI safety in healthcare. This year's iMIMIC was held on September 27th, 2025 in Daejeon, South Korea. There was a very positive response to the call for papers for iMIMIC 2025. We received 22 submissions, of which 15 were accepted as full papers. Each paper was reviewed by at least three reviewers in a double-blind process. The accepted papers focus on introducing the challenges and opportunities related to the topic of interpretability of machine learning systems in the context of medical imaging and computer-assisted intervention. The high quality of the scientific program of iMIMIC 2025 was due first to the authors who submitted excellent contributions and second to the dedicated collaboration of the International Program Committee and the other researchers who reviewed the papers. We would like to thank all the authors for submitting their contributions and for sharing their research activities We are particularly indebted to the Program Committee members and to all the reviewers for their precious evaluations, which permitted us to set up this publication. We were also very pleased to benefit from the participation of the invited speaker Jaesik Choi, Korea Advanced Institute of Science & Technology, South Korea. We would like to express our sincere gratitude to this world-renowned expert.

October 2025

Mauricio Reyes
Wilson Silva
Jaime Cardoso
Pedro Henriques Abreu
José Amorim

iMIMIC 2025 Organization

General Chair

Mauricio Reyes   University of Bern, Switzerland

Program Committee Chairs

Mauricio Reyes	University of Bern, Switzerland
Jaime Cardoso	INESC Porto, Universidade do Porto, Portugal
Pedro Abreu	CISUC and University of Coimbra, Portugal
José Amorim	CISUC and University of Coimbra, Portugal
Wilson Silva	Utrecht University, The Netherlands
Shangqi Gao	University of Cambridge, UK

Steering Committee

Mauricio Reyes	University of Bern, Switzerland
Jaime Cardoso	INESC Porto, Universidade do Porto, Portugal
Jayashree Kalpathy-Cramer	MGH Harvard University, USA
Nguyen Le Minh	Japan Advanced Institute of Science and Technology, Japan
Pedro Abreu	CISUC and University of Coimbra, Portugal
José Amorim	CISUC and University of Coimbra, Portugal
Wilson Silva	Utrecht University, The Netherlands
Mara Graziani	HES-SO Valais-Wallis, Switzerland
Amith Kamath	University of Bern, Switzerland
Hao Chen	Hong Kong University of Science and Technology, China
Shangqi Gao	University of Cambridge, UK

Program Committee

Matan Atad	Technical University of Munich, Germany
Catarina Barata	Instituto Superior Tecnico, Portugal
Alex Bäuerle	Ulm University, Germany

Jaime Cardoso	INESC Porto, Universidade do Porto, Portugal
Pedro Celard	Universidade de Vigo, Spain
Valentina Corbetta	Netherlands Cancer Institute, Netherlands
Ines Domingues	ISEC, Portugal
Bettina Finzel	University of Bamberg, Germany
Tiago Goncalves	INESC TEC, Portugal
Syed Nouman Hasany	University of Rouen, France
Dwarikanath Mahapatra	Inception Institute of Artificial Intelligence, UAE
Fabrice Meriaudeau	University of Burgundy, France
Nataliia Molchanova	Lausanne University Hospital, Switzerland
Helena Montenegro	Universidade do Porto, INESC TEC, Portugal
Axel Mosig	Ruhr University Bochum, Germany
Henning Müller	HES-SO, Switzerland
Angus Nicolson	University of Oxford, UK
Cristiano Patricio	Universidade da Beira Interior, Portugal
Caroline Petitjean	University of Rouen, France
Mauricio Reyes	University of Bern, Switzerland
Tillmann Rheude	Technische Universität Darmstadt, Germany
Isabel Rio-Torto	FEUP, Portugal
Susu Sun	University of Tübingen, Germany
Luis Teixeira	University of Porto, Portugal

Contents

Distribution-Based Masked Medical Vision-Language Model Using Structured Reports

Shreyank N. Gowda[1(✉)], Ruichi Zhang[2], Xiao Gu[3], Ying Weng[4], and Lu Yang[2]

[1] School of Computer Science, University of Nottingham, Nottingham NG8 1BB, UK
shreyank.narayanagowda@nottingham.ac.uk
[2] Department of Computer Science and Technology, School of Informatics, Xiamen University, Xiamen 361005, China
[3] CHI Lab, University of Oxford, Oxford OX3 7DQ, UK
[4] School of Computer Science, University of Nottingham Ningbo China, Ningbo 315100, China

Abstract. Medical image-language pre-training aims to align medical images with clinically relevant text to improve model performance on various downstream tasks. However, existing models often struggle with the variability and ambiguity inherent in medical data, limiting their ability to capture nuanced clinical information and uncertainty. This work introduces an uncertainty-aware medical image-text pre-training model that enhances generalization capabilities in medical image analysis. Building on previous methods and focusing on Chest X-Rays, our approach utilizes structured text reports generated by a large language model (LLM) to augment image data with clinically relevant context. These reports begin with a definition of the disease, followed by the 'appearance' section to highlight critical regions of interest, and finally 'observations' and 'verdicts' that ground model predictions in clinical semantics. By modeling both inter- and intra-modal uncertainty, our framework captures the inherent ambiguity in medical images and text, yielding improved representations and performance on downstream tasks. Our model demonstrates significant advances in medical image-text pre-training, obtaining state-of-the-art performance on multiple downstream tasks.

Keywords: Vision-Language · Uncertainty · Chest X-Ray

1 Introduction

With rapid advancements in deep learning, computer-aided diagnosis in medicine has seen significant progress across various model architectures. However, these models are often trained on specific anatomical or disease categories, requiring expensive data annotation and re-training when applied to new diseases, which limits their broader applicability. Although deep learning has thrived on large-scale labeled datasets from natural images [7], annotating medical images is a

© The Author(s), under exclusive license to Springer Nature Switzerland AG 2026

M. Reyes et al. (Eds.): iMIMIC 2025, LNCS 16464, pp. 1–11, 2026.
https://doi.org/10.1007/978-3-032-17611-0_1

much more time-intensive and costly process. A typical approach involves pre-training on extensive datasets like ImageNet [7] and fine-tuning on specialized medical datasets [27]. However, this method often struggles to achieve generalized performance due to the significant domain gap.

Medical image analysis stands as a critical area in healthcare, where accurate interpretation can significantly impact clinical outcomes. Traditional methods in medical imaging rely heavily on annotated datasets [27], which are costly and time-consuming to curate, especially for new or rare diseases. Recent advances in self-supervised pre-training methods like contrastive predictive coding [22] and masked language modeling [8] have shown promise in leveraging large, unlabeled datasets to learn robust image and text representations. While general vision-language models like CLIP [24] have achieved impressive performance on natural images, they struggle with medical data due to domain-specific language and visual features [10,28]. Existing medical image-text approaches like ConVIRT [29], PRIOR [6], M& M [10] and GLoRIA [14] often overlook the inherent uncertainties present in medical data, where variability in clinical descriptions and visual cues can lead to ambiguous interpretations.

To tackle this, we propose an uncertainty-aware pre-training model for medical image-text data, focusing on X-ray data. We leverage Distribution-based Masked Image-Language Modeling (D-MLM) to capture both inter- and intra-modal uncertainties, thus enabling more nuanced understanding and alignment between images and associated text. By treating representations as probabilistic distributions rather than deterministic points, D-MLM allows the model to capture the natural ambiguity and variability in medical data, enhancing its capacity for accurate and robust prediction. Since existing reports have semantic inconsistencies [10,28], a key component of our approach involves the structured text reports generated by a large language model (LLM) [1]. We first follow M&M [10] that takes the original reports and converts them to a series of 'Observations' and 'Verdicts'. To this, we add at the beginning a definition of the disease, followed by an 'Appearance' section to guide attention to critical regions in the image, and ending with 'Observations' and 'Verdicts' that offer conclusive insights. This structured report provides clinically relevant context that anchors the model's predictions, ensuring that outputs align with medical semantics. We show that using such a structured report significantly improves our overall performance. We use these reports along with their corresponding images to do the pre-training. We show using our approach improves performance on multiple different downstream tasks and different benchmarks.

2   Method

This section presents our uncertainty-aware pre-training framework for medical image-text alignment. Our Distribution-based Masked Image-Language Modeling (D-MLM) approach combines LLM-generated structured reports for clinical context with probabilistic representations that capture inherent medical data ambiguity. Figure 1 provides an overview.

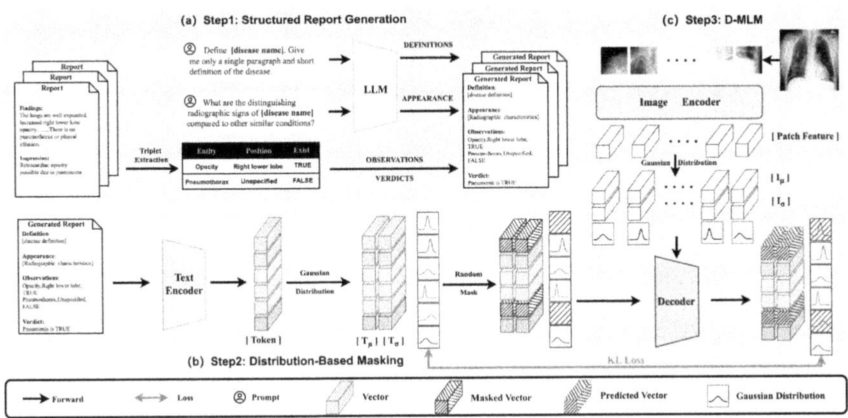

Fig. 1. Overview of our D-MLM framework for medical image-text alignment. (a) Structured Report Generation: An LLM creates standardized reports with disease definitions, appearance guidance, observations, and verdicts. (b) Distribution-Based Representation: Image and text features are encoded as probabilistic distributions with means and variances. (c) D-MLM: The model masks and predicts tokens and patches using distribution-based techniques, optimizing with KL divergence loss.

2.1   Structured Report Generation

We generate structured reports for each medical image using an LLM, following a standardized format with three key components.

Definition: This section provides a concise description of the disease or clinical condition under consideration. To generate this content, we prompt the LLM with, *"Define [disease name]. Give me only a single paragraph and short definition of the disease."* and the model returns a standardized response as, *"Definition: [disease definition]."* This provides a concise introduction to the condition, grounding the model in clinically accurate language.

Appearance: This section describes key radiographic features and diagnostic areas, by prompting the LLM with, *"What are the distinguishing radiographic signs of [disease name] compared to other similar conditions?"* The model then responds with *"Radiographic characteristics: [disease-specific radiographic characteristics]"*, guiding the model's focus to relevant visual cues within the image data.

Observations and Verdicts: This part details visual findings and clinical verdicts, anchoring predictions in medical context. Following Masks & Manuscripts [10], it emphasizes structured clinical reasoning for text-image alignment.

By following this format, the structured reports enhance consistency in text inputs, reduce variability in clinical descriptions, and support the model in achieving precise alignment between image and text features. Details of the definitions and appearances can be found in the link:
https://github.com/kini5gowda/MIMIC-CXR-text

2.2 Distribution-Based Masked Image-Language Modeling (D-MLM)

Our approach centers on Distribution-based Masked Image-Language Modeling (D-MLM), which represents image and text as probabilistic distributions to capture inter-modal and intra-modal uncertainty in medical data.

In D-MLM, image $\mathbf{I}$ is encoded through ImageNet-pretrained ViT-B [9], while text $\mathbf{T}$ uses ClinicalBERT [2]. Both outputs are transformed into multivariate Gaussian distributions, with each token or patch represented as:

$$h_i = N(\mu_i, \sigma_i^2) \tag{1}$$

where μ_i and σ_i^2 are mean and variance vectors. This distribution-based approach captures data variability better than fixed-point representations.

For training, we mask 30% of text tokens (higher than the standard 15% [8]) and focus image masking on diagnostically relevant regions identified from the 'Appearance' section of reports. This adaptive masking strategy emphasizes clinically significant features.

2.3 Pre-training Objective: Distribution-Based Masked Image-Language Modeling

The pre-training objective for D-MLM optimizes the model's ability to reconstruct masked elements using both modalities, framed as a probabilistic reconstruction task. For a masked token or patch h_i, the model predicts:

$$p(h_i | \mathbf{I}, \mathbf{T}_{\setminus i}) = N(\hat{\mu}_i, \hat{\sigma}_i^2) \tag{2}$$

The loss function uses Kullback-Leibler (KL) divergence between predicted and ground truth distributions:

$$\mathcal{L}_{\text{D-MLM}} = \mathbf{E}_{(\mathbf{I},\mathbf{T}) \sim D} \left[\sum_{t \in \mathcal{M}_{\text{text}}} \text{KL} \left(N(\hat{\mu}_t, \hat{\sigma}_t^2) \,\|\, N(\mu_t, \sigma_t^2) \right) \right.$$
$$\left. + \sum_{p \in \mathcal{M}_{\text{image}}} \text{KL} \left(N(\hat{\mu}_p, \hat{\sigma}_p^2) \,\|\, N(\mu_p, \sigma_p^2) \right) \right] \tag{3}$$

where $\mathcal{M}_{\text{text}}$ and $\mathcal{M}_{\text{image}}$ are the sets of masked tokens and patches.

Uncertainty-Aware Alignment Loss. We introduce an alignment loss that minimizes Wasserstein distance between probabilistic embeddings of aligned image-text pairs:

$$\mathcal{L}_{\text{align}} = \sum_{(h_i^{\{\text{text}\}}, h_j^{\{\text{image}\}}) \in \mathcal{A}} W\left(N(\mu_i^{\{\text{text}\}}, \sigma_i^{\{\text{text}\}^2}), \right.$$
$$\left. N(\mu_j^{\{\text{image}\}}, \sigma_j^{\{\text{image}\}^2}) \right) \tag{4}$$

where $\mathcal{A}$ is the set of aligned pairs and $W(\cdot)$ denotes Wasserstein distance. Our ablation studies show this slightly improves performance, though even without it we outperform existing approaches.

Overall Loss Function. The total pre-training loss combines:

$$\mathcal{L}_{\text{total}} = \lambda\mathcal{L}_{\text{D-MLM}} + (1 - \lambda)\mathcal{L}_{\text{align}} \tag{5}$$

where λ balances the contributions of both losses.

3 Experimental Analysis

3.1 Datasets

In this work, we use several publicly available chest X-ray datasets that have been commonly adopted in recent research [6,10,14,28]. **MIMIC-CXR v2** [16] includes 377,110 chest radiographs linked to 227,835 imaging studies, annotated with 14 common chest conditions, which we leverage for pre-training our model. **RSNA Pneumonia Detection** [25] contains approximately 30,000 chest X-rays with bounding box annotations for pneumonia, which we split 60/20/20 for training, validation, and testing. **SIIM-ACR Pneumothorax** [20] comprises 12,954 chest X-rays with image-level pneumothorax annotations and segmentation masks where present; we use a 60/20/20 split for classification tasks and focus on 2,669 samples with pneumothorax for segmentation. **NIH Chest X-Ray Dataset** [26] contains 112,120 frontal-view X-rays from 30,805 patients annotated with 14 thoracic conditions, divided 80/10/10 for training, validation, and testing. **CheXpert** [15] consists of 224,316 X-ray images from 65,240 patients, automatically labeled for 14 thoracic observations, supporting multi-label classification; we split the training data 80/20 and use the official validation set for testing. **COVIDx CXR** [23] includes 29,986 X-rays from 16,648 patients labeled for COVID-19 diagnosis, which we split 70/20/10. **Edema Severity** [4], derived from MIMIC-CXR, comprises 6,524 X-ray images with pulmonary edema severity scores from 0 to 3, which we split 60/20/20 for fine-grained classification.

3.2 Classification

Semi and Fully-Supervised. We conduct both semi-supervised and fully supervised classification experiments across three datasets: RSNA Pneumonia, SIIM-ACR, and CheXpert. The experiments vary the proportion of labeled data from 1% to 100%. For all methods, we report results based on averages and standard deviations over five runs, as provided by PRIOR [6]. The results, presented in Table 1, demonstrate that D-MLM surpasses previous methods by up to 2.3%.

Zero-Shot. We assess zero-shot classification performance of state-of-the-art models on RSNA Pneumonia, SIIM-ACR, and NIH Chest X-Ray datasets, evaluating generalization to 'seen' conditions from different clinical sources. Following MedKLIP [28], we categorize this as zero-shot classification rather than domain adaptation. Table 2 shows our approach outperforming prior methods by up to 1.96% across metrics when evaluated directly after MIMIC-CXR pre-training.

Additionally, we test the model on an entirely new disease, COVID-19, which is absent in the pre-training data. As shown in Table 3, our approach achieves improvements of up to 2.04%.

Table 1. Comparison of semi-supervised and supervised classification results after fine-tuning on RSNA [25], SIIM [20], and CheXpert [15]. Methods are trained on 1%–100% of training data and evaluated using AUC-ROC.

Methods	RSNA Pneumonia			SIIM-ACR			CheXpert		
	1%	10%	100%	1%	10%	100%	1%	10%	100%
MoCo [12]	82.33	85.22	87.90	75.49	81.01	88.43	78.00	86.27	87.24
SimCLR [5]	80.18	84.60	88.07	74.97	83.21	88.72	67.41	86.74	87.97
ConVIRT [29]	83.98	85.62	87.61	84.17	85.66	91.50	85.02	87.58	88.21
GLoRIA [14]	84.12	86.83	89.13	85.05	88.51	92.11	83.61	87.40	88.34
BioViL [3]	81.95	85.37	88.62	79.89	81.62	90.48	80.77	87.56	88.41
LoVT [21]	85.51	86.53	89.27	85.47	88.50	92.16	85.13	88.05	88.27
PRIOR [6]	85.74	87.08	89.22	87.27	89.13	92.39	86.16	88.31	88.61
MedKLIP [28]	87.31	87.99	89.31	85.27	90.71	91.88	86.24	88.14	88.68
M&M [10]	88.11	89.44	91.91	88.81	91.15	93.88	88.45	90.02	90.88
MLIP [18]	89.30	90.04	90.81	–	–	–	89.03	89.44	90.04
UniMedI [13]	90.02	90.41	91.47	–	–	–	89.44	89.72	90.51
IMITATE [19]	91.73	92.85	93.46	–	–	–	89.13	89.49	89.66
D-MLM (Ours)	**91.94**	**92.91**	**93.84**	**91.11**	**92.44**	**95.18**	**89.80**	**90.41**	**91.45**

3.3 Grading

Beyond diagnosis, assessing disease severity is essential. We fine-tune our pre-trained features on the Edema Severity [4] dataset, which classifies conditions on a 0–3 scale. Table 4 shows average scores across all severity levels.

3.4 Segmentation

Table 5 presents our fine-tuning experiments for segmenting three distinct diseases, where we utilize 1%, 10%, and 100% of the available data. Regardless of the varying image distributions associated with each disease, our techniques consistently outperform current leading methods. We see significant gains in particular when data is scarce outperforming previous works by up to 2.16%.

3.5 Implementation Details

To ensure fair comparison, we use a ViT-B [9] image backbone pre-trained on ImageNet [7], with images resized to 224×224, and ClinicalBERT [2] as the text backbone, both with a latent dimension of 768. Training is conducted with a batch size of 128 on 4 NVIDIA Tesla V100 GPUs, using AdamW with a weight decay of 0.05. Definitions and radiographic descriptions are generated by GPT-4 based on specific prompts. In our D-MLM framework, masking ratios are dynamically adjusted: an adaptive ratio for images based on the paired text, and

Table 2. Comparing recent state-of-the-art methods on zero-shot classification task. We use AUC, F1 and ACC scores for comparison. Following MedKLIP [28] for evaluation on NIH Chest X-Ray, the metrics all refer to the macro average on the 14 diseases.

Methods	RSNA Pneumonia			SIIM-ACR			NIH Chest X-Ray		
	AUC↑	F1↑	ACC↑	AUC↑	F1↑	ACC↑	AUC↑	F1↑	ACC↑
ConVIRT [29]	80.42	58.42	76.11	64.31	43.29	57.00	61.01	16.28	71.02
GLoRIA [14]	71.45	49.01	71.29	53.42	38.23	40.47	66.10	17.32	77.00
BioViL [3]	82.80	58.33	76.69	70.79	48.55	69.09	69.12	19.31	79.16
PRIOR [6]	85.58	62.91	77.85	86.62	70.11	84.44	74.51	23.29	84.41
MedKLIP [28]	86.94	63.42	80.02	89.24	68.33	84.28	76.76	25.25	86.19
M&M [10]	88.91	66.58	83.14	91.15	71.58	86.15	77.92	27.55	88.52
D-MLM (Ours)	**90.15**	**68.42**	**85.11**	**91.45**	**72.18**	**86.88**	**79.54**	**28.81**	**90.15**

Table 3. Performance comparison for Zero-Shot Classification on Covid-19 CXR. We use AUC, F1 and ACC scores for comparison.

Methods	AUC↑	F1↑	ACC↑
ConVIRT [29]	52.08	69.02	52.66
GloRIA [14]	66.59	70.07	60.83
BioViL [3]	53.82	69.10	53.75
MedKLIP [28]	73.96	76.70	70.06
M&M [10]	75.15	77.89	73.35
D-MLM (Ours)	**77.19**	**79.52**	**74.78**

a 30% ratio for text to leverage image context. Pre-training is performed for 100 epochs, while fine-tuning occurs over 10 epochs. The learning rate is warmed up to 3×10^{-4} with a cosine scheduler, with encoder rates set to 10^{-5}.

3.6 Ablation Study

We evaluate key components of our approach through focused experiments on the NIH Chest X-Ray dataset in zero-shot settings.

Our structured reports with definition and appearance sections improve performance over M&M [10] and other baselines (Table 6). These sections provide richer clinical context, enhancing image-text alignment. Our D-MLM approach outperforms alternative masking strategies (Table 7), demonstrating the value of modeling features as probabilistic distributions. We use a fixed masking ratio of 0.3 and a λ of 0.2 based on experimental analysis. Notably, even without align-

Table 4. Comparison with state-of-the-art methods on fine-tuning edema severity grading multi-class classification task. Only the average across all classes has been reported here.

Methods	AUC↑	F1↑	ACC↑
ConVIRT [29]	77.00	56.76	69.19
GLoRIA [14]	77.74	57.98	71.45
BioViL [3]	75.40	55.72	69.14
MedKLIP [28]	78.98	58.26	72.80
M&M [10]	80.71	60.18	73.91
D-MLM (Ours)	**82.93**	**62.11**	**75.51**

Table 5. Comparing Dice scores with other state-of-the-art methods on segmentation tasks. We report on three diseases with varying percentages of labeled data 1%, 10%, 100% and see improvements in all cases.

Methods	RSNA Pneumonia			SIIM-ACR			Covid-19		
	1%	10%	100%	1%	10%	100%	1%	10%	100%
Scratch	43.47	60.47	70.68	21.33	33.23	74.47	14.81	23.67	32.28
ConVIRT [29]	57.06	64.91	72.01	54.06	61.21	73.52	19.95	27.24	37.37
GLoRIA [14]	65.55	69.07	73.28	56.73	57.78	76.94	18.89	28.09	38.69
BioVil [3]	68.24	70.38	72.49	62.67	69.98	78.49	21.13	32.39	41.62
PRIOR [6]	70.11	70.88	74.43	66.14	71.24	78.85	23.66	34.72	43.01
MedKLIP [28]	70.64	71.62	75.79	66.59	72.10	79.37	24.45	35.39	43.99
M&M [10]	72.28	73.11	76.68	69.55	73.47	80.28	28.25	37.32	45.04
UniMedI [13]	67.80	73.10	75.30	–	–	–	–	–	–
MLIP [18]	67.70	68.80	73.50	51.60	60.80	68.10	–	–	–
D-MLM (Ours)	**74.11**	**74.77**	**77.16**	**70.95**	**74.49**	**81.12**	**30.41**	**38.11**	**46.11**

ment loss, our model outperforms competitors, indicating that while beneficial, alignment loss is not essential for state-of-the-art performance.

Table 6. Ablation on reports.

Methods	AUC↑	F1↑	ACC↑
Original Report	69.95	20.04	77.71
Triplet	73.48	24.42	82.89
KE-Triplet	76.84	26.11	86.55
M&M [10]	77.92	27.55	88.52
Ours	**79.54**	**28.81**	**90.15**

Table 7. Ablation on masking.

Methods	AUC↑	F1↑	ACC↑
No Masking	61.48	16.33	70.54
MAE [11]	68.84	18.85	75.59
MaskVLM [17]	58.87	14.96	66.69
M&M [10]	77.92	27.55	88.52
D-MLM	**79.54**	**28.81**	**90.15**

4 Conclusion

In this work, we introduced Distribution-based Masked Image-Language Modeling (D-MLM), a novel approach for uncertainty-aware alignment of medical image-text data. By representing both image and text features as probabilistic distributions, D-MLM effectively captures the inherent ambiguity and variability in clinical data, allowing for robust and interpretable multimodal representations. Our method leverages structured reports, dynamically guided masking, and an uncertainty-aware alignment loss to enhance the model's ability to learn meaningful associations between visual and textual information. Extensive experiments demonstrate that D-MLM achieves state-of-the-art performance across multiple medical tasks, highlighting its potential as a foundation for various downstream applications in healthcare.

References

1. Achiam, J., et al.: Gpt-4 technical report. arXiv preprint arXiv:2303.08774 (2023)
2. Alsentzer, E., et al.: Publicly available clinical bert embeddings. arXiv preprint arXiv:1904.03323 (2019)
3. Boecking, B., et al.: Making the most of text semantics to improve biomedical vision–language processing. In: European Conference on Computer Vision, pp. 1–21. Springer (2022)
4. Chauhan, G., et al.: Joint modeling of chest radiographs and radiology reports for pulmonary edema assessment. In: Martel, A.L., et al. (eds.) MICCAI 2020. LNCS, vol. 12262, pp. 529–539. Springer, Cham (2020). https://doi.org/10.1007/978-3-030-59713-9_51

5. Chen, T., Kornblith, S., Norouzi, M., Hinton, G.: A simple framework for contrastive learning of visual representations. In: International Conference on Machine Learning, pp. 1597–1607. PMLR (2020)
6. Cheng, P., Lin, L., Lyu, J., Huang, Y., Luo, W., Tang, X.: Prior: prototype representation joint learning from medical images and reports. In: Proceedings of the IEEE/CVF International Conference on Computer Vision, pp. 21361–21371 (2023)
7. Deng, J., Dong, W., Socher, R., Li, L.J., Li, K., Fei-Fei, L.: Imagenet: a large-scale hierarchical image database. In: IEEE Conference on Computer Vision and Pattern Recognition, pp. 248–255. IEEE (2009)
8. Devlin, J., Chang, M.W., Lee, K., Toutanova, K.: Bert: pre-training of deep bidirectional transformers for language understanding. arXiv preprint arXiv:1810.04805 (2018)
9. Dosovitskiy, A., et al.: An image is worth 16×16 words: transformers for image recognition at scale. arXiv preprint arXiv:2010.11929 (2020)
10. Gowda, S.N., Clifton, D.A.: Masks and manuscripts: advancing medical pre-training with end-to-end masking and narrative structuring. In: International Conference on Medical Image Computing and Computer-Assisted Intervention, pp. 426–436. Springer (2024)
11. He, K., Chen, X., Xie, S., Li, Y., Dollár, P., Girshick, R.: Masked autoencoders are scalable vision learners. In: Proceedings of the IEEE/CVF Conference on Computer Vision and Pattern Recognition, pp. 16000–16009 (2022)
12. He, K., Fan, H., Wu, Y., Xie, S., Girshick, R.: Momentum contrast for unsupervised visual representation learning. In: Proceedings of the IEEE/CVF Conference on Computer Vision and Pattern Recognition, pp. 9729–9738 (2020)
13. He, X., et al.: Unified medical image pre-training in language-guided common semantic space. In: European Conference on Computer Vision, pp. 123–139. Springer (2025)
14. Huang, S.C., Shen, L., Lungren, M.P., Yeung, S.: Gloria: a multimodal global-local representation learning framework for label-efficient medical image recognition. In: Proceedings of the IEEE/CVF International Conference on Computer Vision, pp. 3942–3951 (2021)
15. Irvin, J., et al.: Chexpert: a large chest radiograph dataset with uncertainty labels and expert comparison. In: Proceedings of the AAAI Conference on Artificial Intelligence, vol. 33, pp. 590–597 (2019)
16. Johnson, A.E., et al.: Mimic-cxr, a de-identified publicly available database of chest radiographs with free-text reports. Sci. Data **6**(1), 317 (2019)
17. Kwon, G., Cai, Z., Ravichandran, A., Bas, E., Bhotika, R., Soatto, S.: Masked vision and language modeling for multi-modal representation learning. In: The Eleventh International Conference on Learning Representations (2022)
18. Li, Z., et al.: Mlip: enhancing medical visual representation with divergence encoder and knowledge-guided contrastive learning. In: Proceedings of the IEEE/CVF Conference on Computer Vision and Pattern Recognition, pp. 11704–11714 (2024)
19. Liu, C., Cheng, S., Shi, M., Shah, A., Bai, W., Arcucci, R.: Imitate: clinical prior guided hierarchical vision-language pre-training. IEEE Trans. Med. Imaging (2024)
20. for imaging informatics in medicine, S.: Siim-acr pneumothorax segmentation (2019). https://www.kaggle.com/c/siim-acr-pneumothorax-segmentation
21. Müller, P., Kaissis, G., Zou, C., Rueckert, D.: Joint learning of localized representations from medical images and reports. In: European Conference on Computer Vision, pp. 685–701. Springer (2022)
22. Oord, A.v.d., Li, Y., Vinyals, O.: Representation learning with contrastive predictive coding. arXiv preprint arXiv:1807.03748 (2018)

23. Pavlova, M., et al.: Covid-net cxr-2: an enhanced deep convolutional neural network design for detection of covid-19 cases from chest x-ray images. Front. Med. **9**, 861680 (2022)
24. Radford, A., et al.: Learning transferable visual models from natural language supervision. In: International Conference on Machine Learning, pp. 8748–8763. PMLR (2021)
25. Shih, G., et al.: Augmenting the national institutes of health chest radiograph dataset with expert annotations of possible pneumonia. Radiol. Artif. Intell. **1**(1), e180041 (2019)
26. Wang, X., Peng, Y., Lu, L., Lu, Z., Bagheri, M., Summers, R.M.: Chestx-ray8: hospital-scale chest x-ray database and benchmarks on weakly-supervised classification and localization of common thorax diseases. In: Proceedings of the IEEE Conference on Computer Vision and Pattern Recognition, pp. 2097–2106 (2017)
27. Wen, Y., Chen, L., Deng, Y., Zhou, C.: Rethinking pre-training on medical imaging. J. Vis. Commun. Image Represent. **78**, 103145 (2021)
28. Wu, C., Zhang, X., Zhang, Y., Wang, Y., Xie, W.: Medklip: medical knowledge enhanced language-image pre-training. In: Proceedings of the IEEE/CVF International Conference on Computer Vision (2023)
29. Zhang, Y., Jiang, H., Miura, Y., Manning, C.D., Langlotz, C.P.: Contrastive learning of medical visual representations from paired images and text. In: Machine Learning for Healthcare Conference, pp. 2–25. PMLR (2022)

VLEER: Vision and Language Embeddings for Explainable Whole Slide Image Representation

Anh Tien Nguyen[1], Keunho Byeon[1], Kyungeun Kim[2], and Jin Tae Kwak[1(✉)]

[1] School of Electrical Engineering, Korea University, Seoul 02841, South Korea
`{ngtienanh,bkh5922,jkwak}@korea.ac.kr`
[2] Seegene Medical Foundation, Seoul 04805, South Korea
`kekim23@naver.com`

Abstract. Recent advances in vision-language models (VLMs) have shown remarkable potential in bridging visual and textual modalities. In computational pathology, domain-specific VLMs, which are pre-trained on extensive histopathology image-text datasets, have succeeded in various downstream tasks. However, existing research has primarily focused on the pre-training process and direct applications of VLMs on the patch level, leaving their great potential for whole slide image (WSI) applications unexplored. In this study, we hypothesize that pre-trained VLMs inherently capture informative and interpretable WSI representations through quantitative feature extraction. To validate this hypothesis, we introduce Vision and Language Embeddings for Explainable WSI Representation (*VLEER*), a novel method designed to leverage VLMs for WSI representation. We systematically evaluate *VLEER* on three pathological WSI datasets, proving its better performance in WSI analysis compared to conventional vision features. More importantly, *VLEER* offers the unique advantage of interpretability, enabling direct human-readable insights into the results by leveraging the textual modality for detailed pathology annotations, providing clear reasoning for WSI-level pathology downstream tasks.

Keywords: Computational pathology · Vision-language model · Explainability · Whole slide image

1 Introduction

Recently, there has been growing interest in vision-language models (VLMs), which integrate vision and language modalities by jointly learning from large-scale image-text datasets. A prominent example is CLIP [14], which aligns visual and textual representations through contrastive learning. In computational pathology, this paradigm has been adapted to domain-specific datasets, resulting in pathology VLMs such as PLIP [4], QUILT-Net [5], and CONCH [10]. These models were pre-trained on extensive histopathology datasets containing paired pathology images and descriptive textual data, effectively integrating

© The Author(s), under exclusive license to Springer Nature Switzerland AG 2026
M. Reyes et al. (Eds.): iMIMIC 2025, LNCS 16464, pp. 12–22, 2026.
https://doi.org/10.1007/978-3-032-17611-0_2

visual and textual information in pathology. These VLMs serve as a powerful foundation for downstream tasks in computational pathology. Notably, they have achieved remarkable results in various classification tasks, such as lymph-node metastasis detection, tissue phenotyping, and Gleason grading, often without requiring further training or fine-tuning (i.e., zero-shot learning) [4,5,10]. A close investigation of the existing research on VLMs [4,5,7,10,13,14] reveals that most studies have primarily focused on pre-training VLMs and their direct application to downstream tasks, overlooking two key limitations. First, most prior works mainly focus on patch-level tasks, while WSI-level applications remain largely unexplored. Second, the interpretability of textual embeddings in VLM has not been thoroughly explored, limiting their potential for providing explainable insights in computational pathology.

Herein, we hypothesize that pre-trained VLMs can inherently represent WSIs in a quantitative and interpretable manner, which can be effectively utilized for downstream tasks. To test this hypothesis, we introduce Vision and Language Embeddings for Explainable WSI Representation (*VLEER*), a novel approach for explainable WSI representation. We evaluate *VLEER* on three WSI pathology datasets, systematically proving its effectiveness in downstream tasks. Our experimental results highlight that *VLEER* not only outperforms conventional vision models in WSI analysis but also facilitates direct interpretation of results through human-readable and understandable textual representations, offering a high level of explainability in computational pathology.

2 Methodology

Overall, *VLEER* utilizes two components to learn explainable WSI embeddings: a task-related text pool of pathology keywords and a pre-trained pathology VLM (Fig. 1). The text pool includes a list of pathology keywords that are relevant to downstream tasks, designed to provide explainability via human-readable and understandable pathology terms. To effectively integrate visual (pathology images) and textual (pathology keywords) data, we adopt a pre-trained VLM to align these two modalities by mapping pathology patches of WSIs and their corresponding keywords, generating vision and language patch embeddings. These embeddings are then utilized for downstream analyses. In the following sections, we detail the construction of the text pool, the generation of vision and language embeddings, and the explainability of the proposed embeddings.

2.1 Task-Related Pathology Text Pool

Inspired by [13], we build a text pool of pathology-related keywords. While pathology keywords can be obtained from VLM datasets [4,10], these keywords are generic and not specifically tailored to downstream tasks. Accordingly, we collect task-specific keywords from relevant literature, illustrating the histology of tissues for each task. These keywords include pathological terms that are relevant to both normal and abnormal conditions. For instance, the text pool for the breast cancer subtyping task contains terms related to the histology of invasive ductal carcinoma (IDC), invasive lobular carcinoma (ILC), and normal breast tissues. All collected keywords are then reviewed and validated by a board-certified, experienced pathologist.

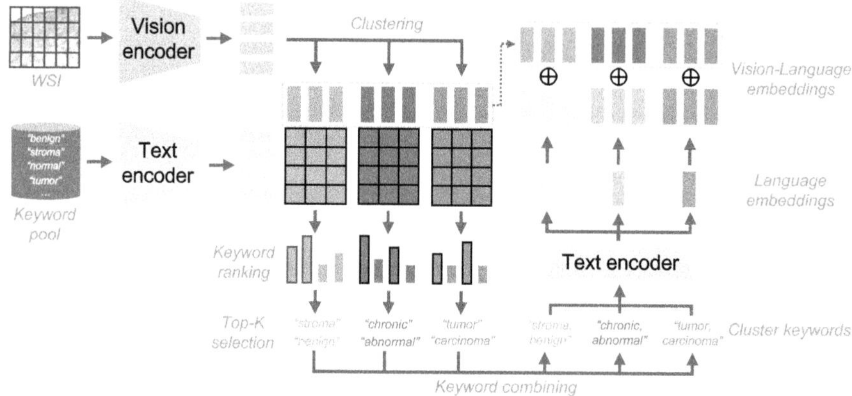

Fig. 1. Vision-language embedding generation in *VLEER*. Tiled patches and curated keywords are embedded using a pre-trained VLM's vision and text encoders. Clustering of vision embeddings, similarities between text and clustered vision embeddings are then measured to select the top-K keywords. These keywords are combined and used to obtain cluster-level language embeddings, which are then concatenated with the corresponding vision embeddings, forming vision-language embeddings.

2.2   Text-Based WSI Embedding

Given a WSI $\mathcal{W}$, a pre-trained VLM $\mathcal{M}$, and a task-related keyword pool $\mathcal{K}$ with N_K keywords, *VLEER* separately processes $\mathcal{W}$ and $\mathcal{K}$ to extract visual and textual embeddings. The two embeddings are then combined to produce a text-based WSI embedding.

Visual Embedding Extraction. $\mathcal{W}$ is tiled into a bag of patches $\mathcal{P} = \{p_i\}_{i=1}^{N_P}$ where N_P is the number of extracted patches. The vision encoder of $\mathcal{M}$ transforms these patches into vision embeddings $\mathcal{V} = \{v_i\}_{i=1}^{N_P}$. Typically, a multiple instance learning (MIL) aggregator is adopted to aggregate these embeddings into a single representative embedding [6,11,15,18]. However, WSIs are inherently *heterogeneous* [2,8,17], and thus conventional approaches may fail to fully capture their complex and diverse characteristics. To address this, we cluster patches into distinct groups to improve the semantic meaning and accuracy of feature representations. For clustering, we employ Lloyd's k-Means method [9] to partition $\mathcal{V}$ into k clusters $\{\mathcal{V}_c\}_{c=1}^{k}$ where $\mathcal{V}_c = \{v_i\}_{i=1}^{N_P^c}$ is the c-th cluster, N_P^c is the number of patches in $\mathcal{V}_c$, and $\sum_{c=1}^{k} N_P^c = N_P$.

Textual Embedding Extraction. We adopt the text encoder of $\mathcal{M}$ to embed all keywords in the pool $\mathcal{K} = \{k_j\}_{j=1}^{N_K}$ into textual embeddings $\mathcal{T} = \{t_j\}_{j=1}^{N_K}$. To enhance the robustness of these text embeddings, we employ various templates to generate diverse text prompts for each keyword, following [10]. The final text embedding t_j for each keyword is then obtained by averaging the embeddings derived from the various text prompts.

Vision-Text Alignment. For each cluster, similarity scores are calculated between all visual and textual embeddings. Mathematically, for the c-th cluster, pair-wise similarity scores are obtained from embeddings in $\mathcal{V}_c$ and $\mathcal{T}$, producing $\mathcal{S}_c = \{s_{i,j} | i = 1, ..., N_P^c \wedge j = 1, ..., N_K\}$ with $s_{i,j}$ is the similarity score between visual embedding $v_i \in \mathcal{V}_c$ and textual embedding $t_j \in \mathcal{T}$.

Cluster Representative Keywords Retrieval. Inspired by [13], we rank all keywords based on their similarity scores with each patch image, aggregate these rankings across all patches within a cluster, and retrieve the most representative keywords for each cluster. Formally, given a visual embedding $v_i \in \mathcal{V}_c$ and textual embeddings $\mathcal{T}$, the ranking of keywords is determined based on $\mathcal{S}_c$ as follows: $\forall (m, n), r_{i,m} < r_{i,n}$ if $s_{i,m} < s_{i,n}$, with $r_{i,j} \in \{1, ..., N_K\}$ is the rank of t_j for v_i. The aggregated rank of each keyword is then calculated by its ranks across all patches in $\mathcal{V}_c$. Finally, we select the top-M keywords with the highest ranks as the cluster representative keywords for the c-th cluster, denoted as $\mathcal{K}_c = \{k_j\}_{j=1}^{M}$.

Vision-Language Embedding Generation. To enhance performance and explainability, we propose integrating cluster representative keywords into the visual embeddings. This is based on the assumption that textual information not only provides semantic interpretability by a direct mapping between images and text keywords, but also enriches the extracted visual embeddings by incorporating complementary insights. Formally, the representative keywords $\mathcal{K}_c = \{k_j\}_{j=1}^{M}$ for the c-th cluster are concatenated by commas and forwarded through the text encoder of $\mathcal{M}$ to obtain the representative textual embedding $\mathcal{T}_c$. Subsequently, we concatenate $\mathcal{T}_c$ with all visual embeddings $v_i \in \mathcal{V}_c$ to produce the vision-language embeddings $\mathcal{V}^* = \{v_i^*\}_{i=1}^{N_P}$. These embeddings are then aggregated into a WSI-level embedding using a trainable MIL aggregator.

2.3 Explainability of Vision and Language Embeddings

Utilizing both vision and language embeddings, *VLEER* offers comprehensive explainability through visual and textual interpretation. Following clustering, adjacent patches within the same cluster are merged into a region of interest (RoI). Each RoI is annotated with the representative keywords. This annotation is region-specific and is generated using Vision-Language embeddings, referred to as a ReVL annotation. We also generate heatmaps using the normalized attention scores from the MIL aggregator to illustrate the contribution of each RoI to the prediction. These complementary visualizations enhance model interpretability by associating key pathology patterns with their predictive significance.

3   Experiments

3.1   Datasets

We evaluate the effectiveness of *VLEER* on cancer subtyping using three public WSI-level datasets from TCGA [1].

Non-Small Cell Lung Carcinoma Subtyping on TCGA-NSCLC. This dataset contains 958 WSIs, with 490 *lung adenocarcinoma* (LUAD) and 468 *lung squamous cell carcinoma* (LUSC) slides. The number of curated keywords for LUAD, LUSC, and normal lung histology are 23, 13, and 20, respectively.

Renal Cell Carcinoma Subtyping on TCGA-RCC. This dataset includes 922 WSIs, consisting of 519 *clear cell renal cell carcinoma* (CCRCC), 294 *papillary renal cell carcinoma* (PRCC), and 109 *chromophobe renal cell carcinoma* (CHRCC) slides. The selected keywords for CCRCC, PRCC, CHRCC, and normal kidney histology are 18, 8, 6, and 31, respectively.

Breast Invasive Carcinoma Subtyping on TCGA-BRCA. This dataset has 1,033 WSIs, with 822 *invasive ductal carcinoma* (IDC) and 211 *invasive lobular carcinoma* (ILC) slides. The collected keywords for IDC, ILC, and normal breast histology are 27, 13, and 26, respectively

Table 1. Comparison between vision (V) and vision-language (V-L) embeddings for cancer sub-typing on three TCGA datasets with four different aggregators. **Bold** numbers indicate higher performance between two types of embeddings.

Aggregator	Emb.	TCGA-NSCLC			TCGA-RCC			TCGA-BRCA			Average		
		Acc	*F*1	*AUC*	*Acc*	*F*1	*AUC*	*Acc*	*F*1	*AUC*	*Acc*	*F*1	*AUC*
ABMIL	V	**0.9035**	**0.9033**	**0.9739**	**0.9425**	**0.9338**	0.9949	0.9313	0.9005	0.9707	0.9257	0.9126	0.9798
	V-L	0.8989	0.8988	0.9679	0.9310	0.9205	**0.9961**	**0.9479**	**0.9225**	**0.9762**	**0.9259**	**0.9139**	**0.9800**
CLAM-SB	V	0.9012	0.9011	**0.9715**	**0.9402**	**0.9282**	0.9951	0.9271	0.8915	0.9649	0.9228	0.9069	0.9771
	V-L	**0.9058**	**0.9056**	0.9696	0.9333	0.9181	**0.9954**	**0.9479**	**0.9196**	**0.9770**	**0.9290**	**0.9144**	**0.9806**
CLAM-MB	V	0.8989	0.8988	0.9691	**0.9333**	**0.9191**	0.9949	0.9292	0.8949	0.9650	0.9205	0.9043	0.9764
	V-L	**0.9103**	**0.9103**	**0.9699**	0.9287	0.9131	**0.9964**	**0.9479**	**0.9196**	**0.9780**	**0.9290**	**0.9143**	**0.9814**
TransMIL	V	0.8736	0.8729	0.9603	0.9333	**0.9247**	0.9882	**0.9146**	**0.8517**	**0.9741**	0.9072	0.8831	0.9759
	V-L	**0.8920**	**0.8918**	**0.9688**	**0.9379**	0.9236	**0.9903**	0.9125	0.8479	0.9728	**0.9141**	**0.8878**	**0.9773**

3.2   Implementation Details

For quantitative evaluation, we compared vision and vision-language embeddings using four MIL aggregators: ABMIL [6], CLAM-SB [11], CLAM-MB [11], and TransMIL [15]. We adopted CONCH [10] as the pre-trained VLM. Three metrics were employed, including accuracy (*Acc*), weighted F1 score (*F*1), and area under the curve (*AUC*). Each experiment was repeated five times with different seeds and reports the average results. All experiments were conducted for 20 epochs using Adam optimizer. The number of clusters and keywords was set to 5. The experiments are conducted using PyTorch with the NVIDIA RTX A6000.

4   Results and Discussion

4.1   Quantitative Evaluation of *VLEER*

Table 1 shows the results of three classification tasks using four different MIL aggregators, comparing vision-only embedding and vision-language embedding. On average, vision-language embeddings consistently achieved higher performance than vision-only embeddings across all evaluation metrics, aggregators, and datasets, with an improvement of $\sim 1\%$ in Acc and F1 score. However, the effect of vision-language embeddings varied among datasets and MIL aggregators. For TCGA-NSCLC, vision-language embeddings showed better performance for all aggregators except ABMIL, with TransMIL showing a substantial improvement of 2.1% in Acc and F1 score. Similarly, for TCGA-BRCA, vision-language embeddings outperformed vision-only embeddings, when paired with ABMIL, CLAM-SB, and CLAM-MB, with a notable gain of 2.3% in Acc and 3.2% in F1 score for CLAM-SB. Regarding TCGA-RCC, vision-language embeddings consistently obtained higher scores for AUC for all aggregators. These findings suggest that while vision-language embeddings enhance WSI classification performance, their effectiveness depends on the datasets and MIL aggregators.

The observed performance improvements suggest that the incorporation of textual features contributes complementary semantic cues that enrich visual representations. Specifically, by embedding pathologically relevant keywords into the vision-language representations, VLEER introduces inductive biases that help disambiguate histologically similar patterns. For instance, visual features alone may struggle to distinguish subtypes with overlapping morphologies, whereas aligned textual cues, such as *hobnailing* or *papillary architecture*, reinforce subtype-specific semantics. This synergy between modalities allows the model to focus on diagnostically salient patterns during aggregation, leading to more robust and discriminative WSI-level embeddings. Hence, the vision-language integration not only supports post-hoc explainability but also directly enhances classification decisions through improved feature expressiveness and contextualization.

4.2   Qualitative Evaluation of *VLEER*

For each WSI in the test set of three datasets (TCGA-BRCA, TCGA-NSCLC, and TCGA-RCC), we generated heatmaps and ReVL annotations for the RoIs. The ReVL annotations were reviewed and validated by a board-certified, experienced pathologist. Overall, the annotations are clinically meaningful and aligned with the histologic patterns observed in the RoIs. Figures 2 and 3 show the heatmaps and ReVL annotations for two representative samples from TCGA-RCC and TCGA-NSCLC, respectively. As illustrated, the ReVL annotations align well with the attention heatmaps, providing a comprehensive and interpretable explanation of the model's predictions.

Figure 2 presents a papillary renal cell carcinoma WSI from TCGA-RCC. The highly attended regions are annotated with *abundant cytoplasm with a reticular*

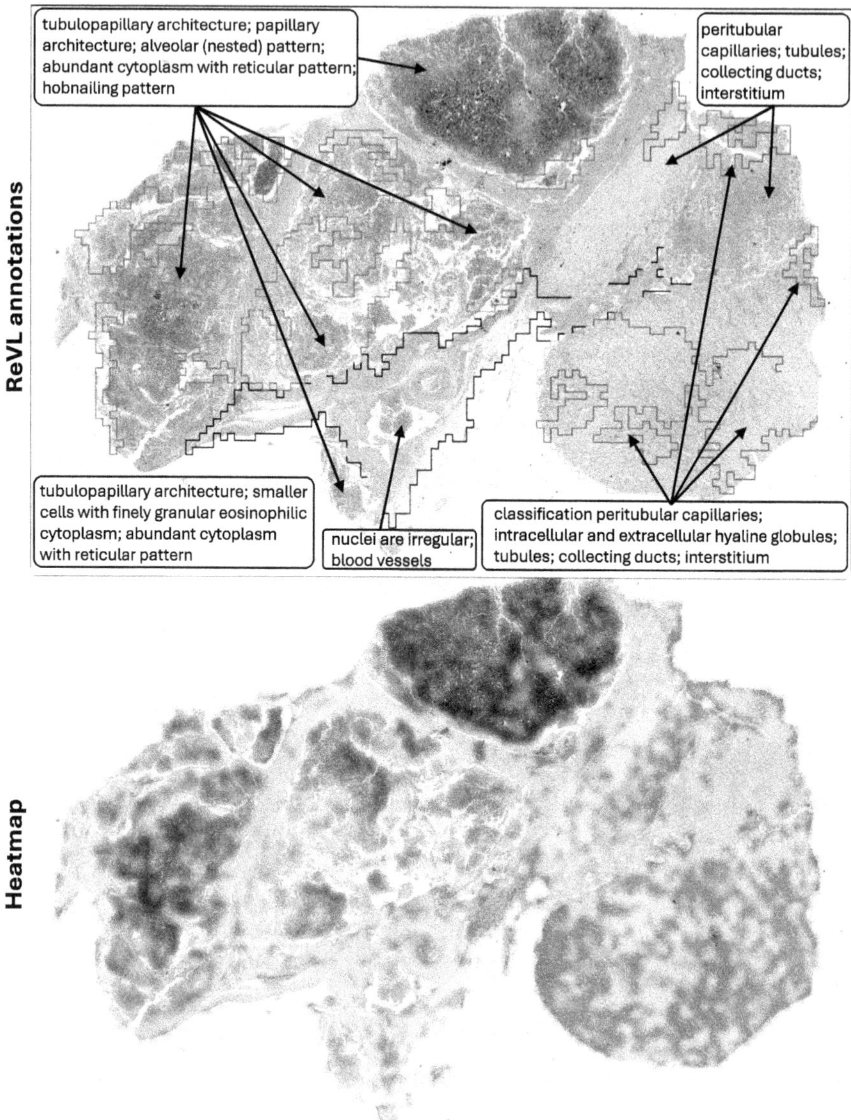

Fig. 2. The ReVL annotations and attention heatmap of a papillary renal cell carcinoma in TCGA-RCC. The highly attended regions (red) in the heatmap are closely related to the patterns of papillary cancer, whereas the low attended regions (green and blue) are normal histology of renal tissues. (Color figure online)

pattern and *hobnailing pattern*, which are key pathological features of papillary carcinoma. In contrast, normal regions, annotated as *peritubular capillaries* and *tubules*, contribute minimally to cancer subtyping predictions. Figure 3 shows a

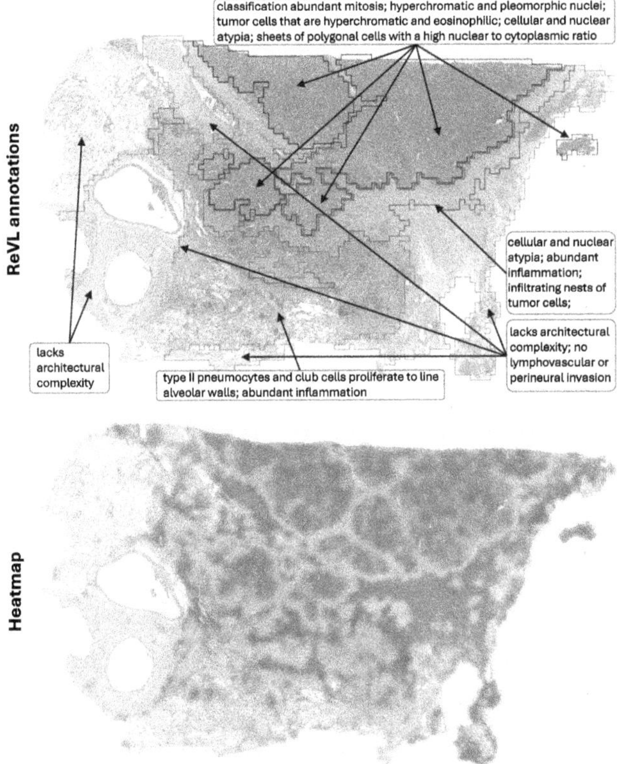

Fig. 3. The ReVL annotations and attention heatmap of a lung squamous cell carcinoma in TCGA-NSCLC. The highly attended regions (red) in the heatmap are closely related to the patterns of squamous cancer, whereas the low attended regions (green and blue) are normal histology of lung tissues. (Color figure online)

lung squamous cell carcinoma WSI from TCGA-NSCLC. The ReVL annotations capture cell-level patterns of cancerous regions, highlighting key histopathology features such as *abundant mitosis* and *cellular and nuclear atypia*. Non-tumor regions receive lower attention, with characteristics such as *a lack of architectural complexity* and *no lymphovascular or perineural invasion*. In summary, the combination of ReVL annotations and heatmaps enables the precise identification of abnormal regions within a WSI along with their pathological patterns. This approach enhances explainability by linking the model's predictions to highly attended RoIs and their corresponding pathology characteristics.

4.3   Explainability of *VLEER*

The level of explainability is particularly significant in computational pathology, where deep learning models often perceived as black boxes, making it difficult for clinicians and pathologists to fully trust their outputs [12,13]. Unlike traditional methods, *VLEER* enhances transparency by providing text-based justifications that align with established pathology knowledge. This improved transparency and interpretability are essential for clinical adoption, as medical professionals require not only accurate predictions but also a clear rationale behind the models' decisions. By bridging the gap between machine learning and human expertise, *VLEER* facilitates a more intuitive understanding of deep learning-based analyses, fostering greater confidence and acceptance in real-world diagnostic settings.

Furthermore, this interpretability extends beyond simple classification by enabling fine-grained insights into the spatial organization of histological structures. By clustering patches and assigning pathology annotations, *VLEER* preserves the contextual relationships within WSIs. This capability is particularly valuable in tasks such as tumor subtyping, cancer grading, and biomarker prediction, where localized histological variations can significantly impact clinical decisions [3, 16]. Moreover, the ability to highlight RoIs alongside their corresponding pathological patterns allows pathologists to validate models' predictions.

We further examined misclassified WSIs to evaluate whether the ReVL annotations remain histologically meaningful even when the final prediction is incorrect. In several such cases, we observed that ReVL annotations often still captured histologically valid patterns, though sometimes more aligned with the incorrect class. This indicates that while our method highlights meaningful regions, classification errors may stem from ambiguous features or limitations in the aggregation process. Additionally, the approach relies on the quality of the keyword pool and the alignment accuracy of the pre-trained VLM, which may be less effective for rare subtypes or atypical morphologies. These factors represent key limitations to be addressed in future work.

5   Conclusion

Herein, we propose *VLEER* for generating vision-language embeddings in WSI representation. The method not only increases the performance on downstream tasks, but also provides explainability of the prediction. The combination of ReVL annotations and attention heatmaps forms a powerful and interpretable framework for WSI analysis. By integrating pathologically meaningful features with spatially aware visual attention, our approach enhances both model performance and transparency, making AI-driven pathology workflows more interpretable and clinically applicable by enabling clinicians to better understand models' decision-making process.

Acknowledgement. This work was supported by the National Research Foundation of Korea (NRF) (No. RS-2025-00558322).

References

1. The Cancer Genome Atlas Program (TCGA) — cancer.gov. https://www.cancer.gov/ccg/research/genome-sequencing/tcga
2. Chan, T.H., Cendra, F.J., Ma, L., Yin, G., Yu, L.: Histopathology whole slide image analysis with heterogeneous graph representation learning. In: Proceedings of the IEEE/CVF Conference on Computer Vision and Pattern Recognition, pp. 15661–15670 (2023)
3. Fuchs, T.J., Buhmann, J.M.: Computational pathology: challenges and promises for tissue analysis. Comput. Med. Imaging Graph. **35**(7–8), 515–530 (2011)
4. Huang, Z., Bianchi, F., Yuksekgonul, M., Montine, T.J., Zou, J.: A visual–language foundation model for pathology image analysis using medical twitter. Nat. Med. **29**(9), 2307–2316 (2023). https://doi.org/10.1038/s41591-023-02504-3, https://doi.org/10.1038/s41591-023-02504-3
5. Ikezogwo, W.O., et al.: Quilt-1m: one million image-text pairs for histopathology. In: Thirty-seventh Conference on Neural Information Processing Systems Datasets and Benchmarks Track (2023). https://openreview.net/forum?id=OL2JQoO0kq
6. Ilse, M., Tomczak, J., Welling, M.: Attention-based deep multiple instance learning. In: International Conference on Machine Learning, pp. 2127–2136. PMLR (2018)
7. Jia, C., et al.: Scaling up visual and vision-language representation learning with noisy text supervision. In: Meila, M., Zhang, T. (eds.) Proceedings of the 38th International Conference on Machine Learning. Proceedings of Machine Learning Research, vol. 139, pp. 4904–4916. PMLR, 18–24 July 2021. https://proceedings.mlr.press/v139/jia21b.html
8. Levy-Jurgenson, A., Tekpli, X., Kristensen, V.N., Yakhini, Z.: Spatial transcriptomics inferred from pathology whole-slide images links tumor heterogeneity to survival in breast and lung cancer. Sci. Rep. **10**(1), 18802 (2020)
9. Lloyd, S.: Least squares quantization in pcm. IEEE Trans. Inf. Theory **28**(2), 129–137 (1982). https://doi.org/10.1109/TIT.1982.1056489
10. Lu, M.Y., et al.: A visual-language foundation model for computational pathology. Nat. Med. **30**, 863–874 (2024)
11. Lu, M.Y., Williamson, D.F., Chen, T.Y., Chen, R.J., Barbieri, M., Mahmood, F.: Data-efficient and weakly supervised computational pathology on whole-slide images. Nat. Biomed. Eng. **5**(6), 555–570 (2021)
12. Nguyen, A.T., Kwak, J.T.: Gpc: generative and general pathology image classifier. In: International Conference on Medical Image Computing and Computer-Assisted Intervention, pp. 203–212. Springer (2023)
13. Nguyen, A.T., Vuong, T.T.L., Kwak, J.T.: Towards a text-based quantitative and explainable histopathology image analysis. In: International Conference on Medical Image Computing and Computer-Assisted Intervention, pp. 514–524. Springer (2024)
14. Radford, A., et al.: Learning transferable visual models from natural language supervision. In: Meila, M., Zhang, T. (eds.) Proceedings of the 38th International Conference on Machine Learning. Proceedings of Machine Learning Research, vol. 139, pp. 8748–8763. PMLR, 18–24 July 2021. https://proceedings.mlr.press/v139/radford21a.html
15. Shao, Z., Bian, H., Chen, Y., Wang, Y., Zhang, J., Ji, X., et al.: Transmil: transformer based correlated multiple instance learning for whole slide image classification. Adv. Neural. Inf. Process. Syst. **34**, 2136–2147 (2021)

16. Verghese, G., et al.: Computational pathology in cancer diagnosis, prognosis, and prediction-present day and prospects. J. Pathol. **260**(5), 551–563 (2023)
17. Wu, J., Ke, X., Jiang, X., Wu, H., Kong, Y., Shao, L.: Leveraging tumor heterogeneity: heterogeneous graph representation learning for cancer survival prediction in whole slide images. Adv. Neural. Inf. Process. Syst. **37**, 64312–64337 (2025)
18. Zhang, J., Nguyen, A.T., Han, X., Trinh, V.Q.H., Qin, H., Samaras, D., Hosseini, M.S.: 2dmamba: efficient state space model for image representation with applications on giga-pixel whole slide image classification (2024). https://arxiv.org/abs/2412.00678

Do Segmentation Models Understand Vascular Structure? A Blob-Based Saliency Framework

Guillaume Garret[1(✉)], Antoine Vacavant[1] , and Carole Frindel[2]

[1] Université Clermont Auvergne, CHU Clermont-Ferrand, Clermont Auvergne INP, CNRS, Institut Pascal, 63000 Clermont-Ferrand, France
`Guillaume.GARRET@uca.fr`
[2] Laboratoire CREATIS, Université Lyon I, Université Claude Bernard, CNRS, Inserm, INSA-Lyon, Lyon, France

Abstract. Deep learning models achieve strong performance in vascular segmentation but often lack transparency, a critical factor for clinical adoption. We introduce a novel explainability framework combining graph-based point selection and blob-level analysis of saliency maps to interpret model behavior. By linking anatomically relevant points from the vascular graph to localized attributions, we evaluate how model predictions align with domain-specific features such as tubularity, connectivity, and thickness. Our analysis on two vascular datasets reveals that model decisions are primarily driven by local visual cues, with limited evidence of global anatomical reasoning. These findings highlight the need for structured explainability tools and architectures that better integrate anatomical context in medical image segmentation.

Keywords: deep learning · explainable artificial intelligence · attribution · vessel segmentation

1 Introduction

Deep learning is now gold standard in medical tasks like vascular segmentation, but improving accuracy often requires more complex models. Techniques such as ensembling [9,17], attention modules [12,19], and joint models [14] enhance segmentation quality but reduces interpretability.

At the same time, interpretable AI is seen as the main concern of practitioners [4]. Ensuring that the basis of models' decisions aligns with domain expert knowledge is then crucial. In particular, a reliable vessel segmentation should rely on both local features, like shape and intensity, as well as global features, such as vessel inter-connections, which are meaningful in a broader context.

Attribution methods [1,2,11] offer a way to understand model decisions by identifying the most influential input features, *e.g.* voxels, in the model's decision [13,18]. Although rarely used in segmentation, attribution can also be valuable

© The Author(s), under exclusive license to Springer Nature Switzerland AG 2026
M. Reyes et al. (Eds.): iMIMIC 2025, LNCS 16464, pp. 23–32, 2026.
https://doi.org/10.1007/978-3-032-17611-0_3

in this context, considering segmentation as voxel-wise classification [13]. We hypothesize that **1.** Positive attribution values should be concentrated around vessel voxels and their immediate neighborhoods, while surrounding tissue voxels should yield near-zero or negative attribution values. **2.** The spatial location of positive attribution regions should closely align with vascular pathways, indicating that the classification of a voxel depends on its belonging to a broader vascular object. **3.** The patterns of attribution should vary depending on vascular characteristics (*e.g.* vessel size, branching topology, or connectivity) indicating the model's sensitivity to different anatomical features, which embody various visual vessel patterns.

On two vascular datasets, we extract ground-truth graphs and define anatomically meaningful point of interest (POIs), such as bifurcations, endpoints and mid-branches. We compute gradient-based attribution maps with respect to these POIs and introduced a blob-based approach to detect and characterize influential regions. By comparing attribution blob characteristics to vessel features like connectivity and thickness, we quantify impact of such domain-knowledge on model decisions.

2 Methods

To test our hypotheses, we propose a knowledge-guided analysis framework (see Fig. 1). First, we extract the vessel graph from the segmentation ground truth and define POIs at key locations such as bifurcations, endpoints, and segments (Sect. 2.1). We compute attribution maps w.r.t. each POI (Sect. 2.2), apply blob detection to identify salient attribution regions (Sect. 2.3), and then correlate blobs characteristics with the vascular properties of the POI (Sect. 2.4).

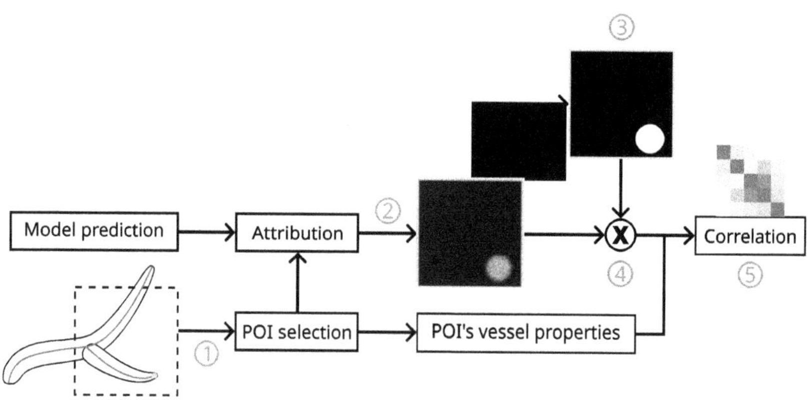

Fig. 1. Framework overview. (1) POIs (green) are extracted from the vessel graph (red). (2) Saliency maps are computed and multiplied (4) by detected blob masks (3). (5) The masked attributions are analyzed in relation to the POIs' vascular properties to assess anatomical relevance in model decisions. (Color figure online)

2.1 Vascular Points of Interest

In the following, "voxel" refers to an input voxel, while "logit" refers to an output voxel. Gradient-based attribution methods are logit-specific and do not provide class-based explanation, especially in label-encoded segmentation tasks. By considering segmentation as a voxel-wise classification problem, we propose to individually explain a subset of logits (POIs) selected based on their anatomical relevance within the vascular structure.

Vascular networks are tree-like topology that can be modeled by a hierarchical graph, as illustrated in Figs. 2a and 2c. Vessel recognition should, at least in part, depend on global anatomical context, such as connections between branches. To capture this, we extract the vascular graph from the segmentation ground truth and define POIs based on its nodes and edges. Nodes correspond to anatomical landmarks such as bifurcations and endpoints. Edges, represented by centerline midpoints, describe regular tubular vessel segments.

2.2 Voxel-Wise Attribution

We use saliency maps [15] as a baseline attribution method in this study. Saliency is a simple and widely used gradient-based approach that highlights which input voxels most influence the model's prediction. Formally, let F be the model's output (logit) given the input image x. The saliency value S of a voxel i is defined as the partial derivative of the logit w.r.t. the input intensity $x_i : S_i(x) = \frac{\partial F(x)}{\partial x_i}$.

The sign of an attribution indicates the direction of influence: in binary label-encoded segmentation, a positive attribution means the voxel contributes toward the "vessel" class (label 1), while a negative attribution indicates influence toward the "background" class (label 0). In our case, $F(x)$ corresponds to the output value of a logit from our set of POIs. Hence, analyzing the attribution patterns around these POIs allows to investigate whether the vascular structure influences the model's predictions.

2.3 Attribution Blob Detection

The model's decision-making is reflected in the saliency maps. A qualitative inspection reveals that high attribution values are concentrated in localized, blob-like regions, while most voxels have near-zero attributions. To mitigate the dilution effect of these low-attribution voxels, we focus our analysis on the attribution blobs, *i.e.* regions where saliency is concentrated.

Assuming Hypothesis 2 (positive attributions should align with vascular pathways) we developed a custom blob detector combining the multiscale Frangi filter [6] and Otsu thresholding.

The Frangi filter enhances curvilinear and blob-like structures in the saliency maps, capturing the diverse shapes of vascular networks. We focused on white ridges, corresponding to regions of positive attributions that indicate support for the "vessel" class. Frangi's scale-space parameter was set to $\sigma \in [2, 16]$ to detect both fine and coarse structures. The shape sensitivity parameters were set to $\alpha = 0.5$, $\beta = 0.5$, and $C = 15$, chosen empirically to balance detection of elongated vessels and compact blobs.

After applying the Frangi filter, we used the Otsu algorithm (with 256 bins) to threshold the enhanced saliency maps into binary masks, separating blobs from the background. This combination of Frangi filtering (for geometric enhancement) and Otsu thresholding (for segmentation) enables robust detection of attribution blobs across varying scales and shapes. The parameters were selected empirically based on visual inspection of attribution distributions and produced consistent, satisfactory results across datasets.

2.4 Domain-Specific Features

We consider a set of vascular and image-based features to help interpret model behavior. Vascular features include connectivity, thickness, volume, and tubularity. Connectivity is derived from the graph, thickness is computed via Euclidean distance transform, and tubularity leverages second-order image information. Image features include the POI voxel intensity, the number and volume of vessels within the patch, and the distance from the POI to the patch center. Some features, like vessel connectivity, are adapted from the image to the patch scale to match our patch-based training [8].

These features are later analyzed in the Results to explore their relationship with the attribution blobs.

3 Experimental Setup

3.1 Datasets

IRCAD. [16] is a public dataset[1] of 20 3D chest CT scans (10 males and 10 females), including liver vessels. We adopt a 5-fold cross-validation scheme: each fold uses 16 images for training, 3 for validation, and leaves out 1 image for testing. For each test image, we extract vascular structures (Sect. 2.1), yielding a total of 2,501 vascular POIs for our explainability analysis. This dataset is particularly valuable for explainability studies due to its diversity (organ variability, presence of tumors in 75% of cases, metal artifacts, *etc.*), but its annotations are known to be imperfect [10]. To mitigate this limitation, we complement it with the Bullitt dataset.

Bullitt. [3] is a cerebral MRA dataset containing 100 images from healthy volunteers, evenly distributed according to age and sex. Per-voxel annotations are available for 33 volumes, as provided by the community[2] [10]. Unlike IRCAD, the test set for Bullitt is defined independently of the folds and consists of 5 images. Vascular graph extraction on these test images yields a total of 6,243 POIs for analysis. The remaining images are split into training and validation sets using an 80:20 ratio, applied in a 5-fold cross-validation scheme.

---

[1] Link to a downloadable pre-processed version: http://eidolon.univ-lyon2.fr/~jlamy.
[2] See footnote 1.

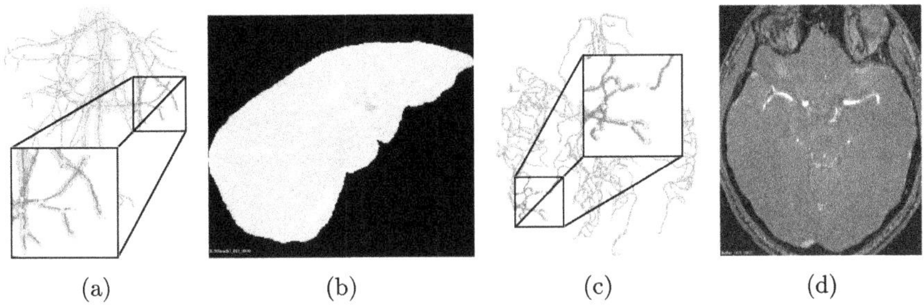

Fig. 2. Vessel graph and slice example for IRCAD (a and b) and Bullitt (c and d)

3.2   Model

We selected a U-Net architecture for this study, as it remains a widely used and state-of-the-art baseline in vascular segmentation tasks. U-Net and its variants continue to serve as a standard point of comparison in recent works [5,19].

Since model optimization is not the primary focus of this work, we used hyperparameters reported in the literature [7] for training. Our models achieve Dice scores of $DSC = 0.63 \pm 0.043$ and $DSC = 0.64 \pm 0.0035$ for IRCAD and Bullitt, respectively. For Bullitt, we analyze the best model from the 5-fold cross-validation. For IRCAD, we explain each model M_k with respect to its corresponding test image I_k, following the leave-one-out strategy.

3.3   Evaluation

We evaluate the attribution maps along two axes: **1.** the contrast between blobs and background, and **2.** the relationship between blob locations and anatomical POIs.

To assess the relative influence of voxels within blob regions compared to the background, we compute two contrast measures. The Fisher contrast-to-noise ratio (FCNR, Eq. (1a)) compares the mean attribution values μ between blobs and background, accounting for variability via the attribution variance σ. A higher FCNR indicates more focused attributions within the blob. The L1-norm ratio (Eq. (1b)) compares the total absolute attribution values between the two regions, where S_{blob} and S_{bg} denote the attribution values within the blob and background areas respectively. A higher ratio suggests stronger localization of attributions within the blob.

$$\text{FCNR} = \frac{(\mu_{\text{blob}} - \mu_{\text{bg}})^2}{\sigma^2_{\text{blob}} + \sigma^2_{\text{bg}}} \qquad (1a) \qquad\qquad R_{\|\cdot\|_1} = \frac{\|S_{\text{blobs}}\|_1}{\|S_{\text{bg}}\|_1} \qquad (1b)$$

To investigate the spatial alignment between the model's attributions and the vascular anatomy, we compute the Euclidean distance between each explained

POI and the centroid of the corresponding attribution blob detected in the saliency map.

Finally, we analyze how vascular characteristics influence the model's decision-making by computing Spearman's rank correlation coefficient (Eq. (2)) between the properties of attribution blobs and the corresponding vascular features of the POIs. This non-parametric measure quantifies the strength of monotonic relationships between two variables x and y, computed as:

$$r_s = \frac{\mathrm{cov}(r_g(x), r_g(y))}{\sigma_{r_g(x)}\sigma_{r_g(y)}} \tag{2}$$

where $r_g(x)$ and $r_g(y)$ denote the ranks of the variables.

4   Results

The following results focus on the analysis of well-classified points of interest (True Positives), based on $2,612$ saliency maps for IRCAD and $8,529$ for Bullitt.

4.1   Blob-Like Attribution

Our blob detector reveals that 99.7% of IRCAD cases and 94.6% of Bullitt cases exhibit compact, highly contrasted attribution blobs in the saliency maps, as shown in Fig. 3.

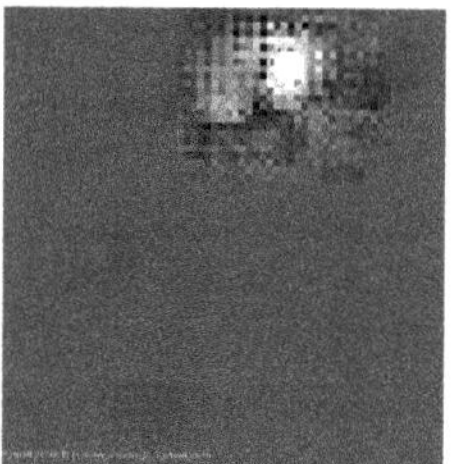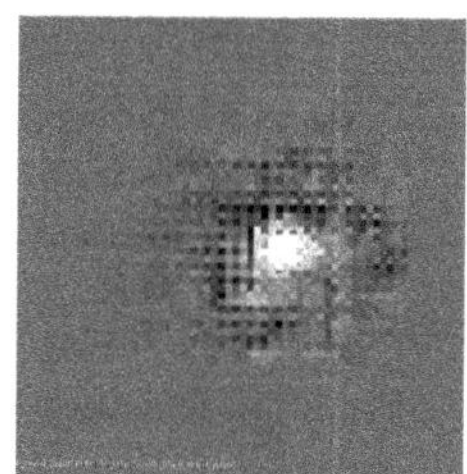

Fig. 3. 2D slices of saliency maps with detected blobs delimited in green. (Color figure online)

The FCNR for these blobs is 0.78 ± 0.35 in IRCAD and 1.5 ± 0.71 in Bullitt. The L1-norm ratio further indicates that the average influence of a voxel within a blob is 109 ± 53.5 times higher for IRCAD and 356 ± 199 times higher for Bullitt compared to background voxels. These metrics quantify the distinctiveness and relative importance of blob regions in the model's output. As shown in Fig. 4, the majority of explanatory power in the saliency maps is concentrated within the blob regions, highlighting the model's strong focus on localized features.

We also examined the spatial alignment between blobs and their associated POIs by measuring the Euclidean distance between each blob centroid and its

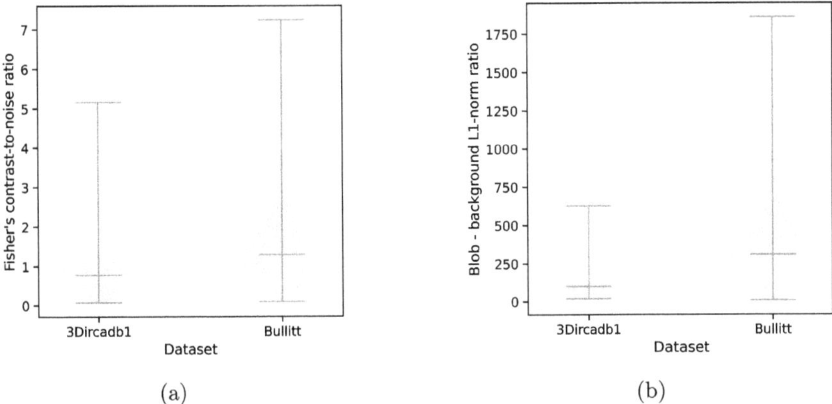

Fig. 4. Violin plot of (a) the Fisher contrast-to-noise ratio and (b) the average L1-norm ratio between attribution within blob and background regions.

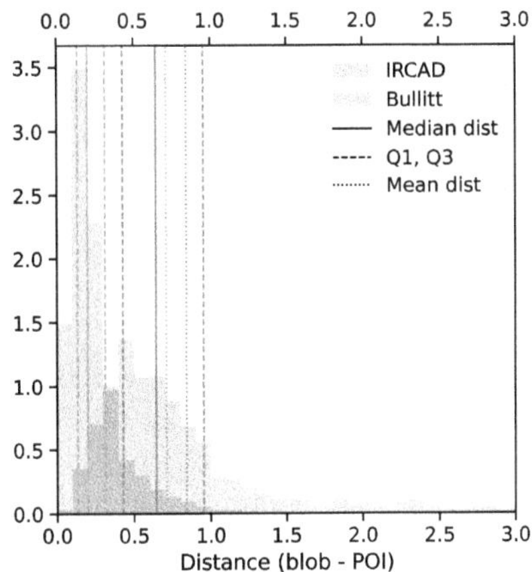

Fig. 5. Histograms of distances between detected blobs' centroids and POIs.

corresponding POI. The distance distributions (see Fig. 5) are heavily skewed toward low values. The median distances are 0.64 voxels for IRCAD and 0.20 voxels for Bullitt, with the third quartile remaining below 1 voxel in both datasets. These results confirm that the model's most influential features are located in the immediate neighborhood of the predicted point.

4.2   Vascular Features Influence

The Spearman's rank correlations coefficients between domain-specific vascular features and attribution descriptors reveal generally weak correlations in the IRCAD dataset, and slightly stronger but still modest correlations in Bullitt. The correlation matrices are given in Fig. 6.

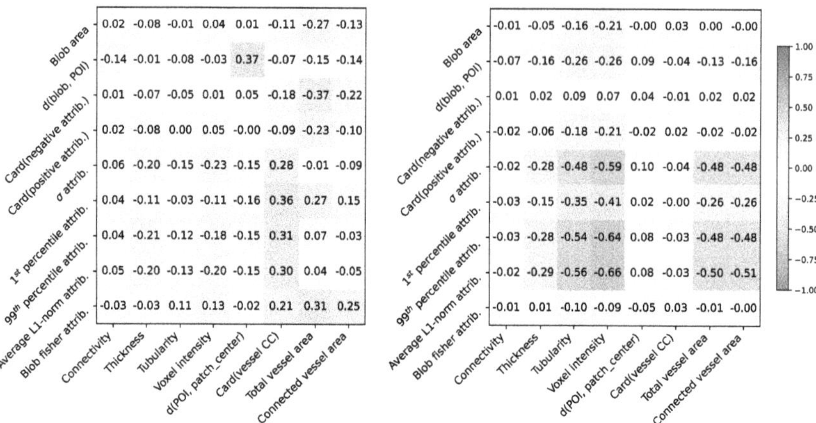

Fig. 6. Spearman's rank correlations for IRCAD's (left) and Bullitt's (right) saliency maps.

These results suggest that the model predominantly relies on local visual cues, rather than integrating higher-level vascular structure information, to make its predictions.

5 Discussion

Our results highlight the critical role of the immediate neighborhood in U-Net predictions, partially confirming Hypothesis 1. The model's focus is localized around predicted points, showing a heavy reliance on local visual cues for vascular segmentation. In contrast, the presence of blob-like positive attribution regions contradicts Hypothesis 2, which expected attributions to follow vessel paths. Instead, the model produces compact attribution blobs, suggesting limited use of global vascular structures. This raises questions about the effective receptive field of the model. Despite the established benefits of large patches in training, as highlighted in the literature, our results imply that a limited voxel subset drives the predictions. This inconsistency calls for further investigation, including an analysis of blob variations across different patch sizes.

Weak correlations between attribution descriptors and vascular features disconfirm Hypothesis 3. The model's attributions do not appear influenced by properties such as connectivity, thickness, or tubularity, although these properties reflect various visual patterns of vessels.

Our findings also highlight the dependence of results on the blob detection pipeline (Frangi + Otsu). While empirically tuned parameters produced consistent results, a more systematic optimization is needed to ensure robustness across anatomical contexts. Interestingly, positive correlations between attributions strength and contrast, and the quantity of vessels in the patch appears in IRCAD but not in Bullitt, reflecting possible differences in blob detection performance due to vessel size and anatomical context. Liver vessels (IRCAD), being larger, better match the blob detection scales, while brain vessels (Bullitt) may be missed or produce diffuse blobs.

Overall, the model's attributions are primarily driven by local intensity patterns, with limited evidence of global anatomical reasoning. Weak correlations with vascular features suggest that the model does not fully leverage anatomical properties such as connectivity, thickness, or tubularity. This highlights a limitation of standard convolutional architectures trained with voxel-wise supervision: they excel at capturing local details but may overlook the broader anatomical context. Preliminary analyses comparing blob sizes to vessel size distributions across datasets suggest that the model may adapt to global dataset characteristics, such as typical vessel size, rather than relying on local anatomical features. This hypothesis will be further explored in future work.

6   Conclusion

We introduced an explainability framework for vessel segmentation models that links POIs from the vascular graph to localized attributions in saliency maps, enabling a fine-grained, anatomically informed evaluation of model behavior. Our results show that model predictions are primarily driven by local visual cues, with limited use of vascular features such as connectivity, bifurcations, or topology — features that are essential for clinical reasoning but remain underutilized.

This highlights a key limitation of standard convolutional architectures: while effective locally, they tend to overlook the broader anatomical context required for robust and interpretable vascular segmentation. Future work should explore the integration of topological priors, graph-based reasoning, or structural constraints to guide models beyond local texture recognition toward anatomically grounded understanding. While demonstrated here for vascular segmentation, our framework is broadly applicable to any task involving tree-like or curvilinear structures, such as pulmonary airways or road networks.

Acknowledgments. This project receives support from the Région Auvergne-Rhône-Alpes under the DAISIES project. We would like to thank the NVIDIA Academic Hardware Grant Program for providing the GPU resources used in this work.

Compliance with Ethical Standards. Ethical approval was not required as confirmed by the license attached with the open access data.

Disclosure of Interests. The authors have no competing interests to declare.

References

1. Ancona, M., Ceolini, E., Öztireli, C., Gross, M.: Towards better understanding of gradient-based attribution methods for deep neural networks. In: ICLR (2018)

2. Ancona, M., Öztireli, C., Gross, M.: Explaining deep neural networks with a polynomial time algorithm for shapley value approximation. In: ICML, vol. 97, pp. 272–281 (2019)

3. Bullitt, E., et al.: Vessel tortuosity and brain tumor malignancy: a blinded study1. Acad. Radiol. **12**, 1232–1240 (2005)

4. Chouvarda, I., et al.: Differences in technical and clinical perspectives on ai validation in cancer imaging: mind the gap! Eur. Radio. Exp. p. 7 (2025)

5. Dang, V.N., et al.: Vessel-captcha: an efficient learning framework for vessel annotation and segmentation. Med. Image Anal. **75**, 102263 (2022)

6. Frangi, A.F., Niessen, W.J., Vincken, K.L., Viergever, M.A.: Multiscale vessel enhancement filtering. In: MICCAI, pp. 130–137 (1998)

7. Garret, G., Vacavant, A., Frindel, C.: Deep vessel segmentation based on a new combination of vesselness filters. In: IEEE ISBI, pp. 1–5 (2024)

8. Garret, G., Vacavant, A., Frindel, C.: Do segmentation models understand vascular structure? A blob-based XAI framework (2025)

9. Isensee, F., Jaeger, P.F., Kohl, S.A.A., Petersen, J., Maier-Hein, K.H.: nnU-net: a self-configuring method for deep learning-based biomedical image segmentation. Nat. Methods 203–211 (2021)

10. Lamy, J., Merveille, O., Kerautret, B., Passat, N.: A benchmark framework for multiregion analysis of vesselness filters. IEEE Trans. Med. Imaging **41**(12), 3649–3662 (2022)

11. Lundberg, S.M., Lee, S.I.: A unified approach to interpreting model predictions. In: NeurIPS, vol. 30 (2017)

12. Oktay, O., et al.: Attention u-net: learning where to look for the pancreas. In: MIDL (2018)

13. Saleem, H., Raza Shahid, A., Raza, B.: Visual interpretability in 3D brain tumor segmentation network. Comput. Biol. Med. **133**, 104410 (2021)

14. Shin, S.Y., Lee, S., Yun, I.D., Lee, K.M.: Deep vessel segmentation by learning graphical connectivity. Med. Image Anal. **58**, 101556 (2019)

15. Simonyan, K., Vedaldi, A., Zisserman, A.: Deep inside convolutional networks: visualising image classification models and saliency maps. In: ICLR (2014)

16. Soler, L., et al.: 3d image reconstruction for comparison of algorithm database (2010)

17. Survarachakan, S., Pelanis, E., Khan, Z.A., Kumar, R.P., Edwin, B., Lindseth, F.: Effects of enhancement on deep learning based hepatic vessel segmentation. Electronics **10**(10) (2021)

18. Wargnier-Dauchelle, V., Grenier, T., Durand-Dubief, F., Cotton, F., Sdika, M.: A weakly supervised gradient attribution constraint for interpretable classification and anomaly detection. IEEE Trans. Med. Imaging **42**, 3336–3347 (2023)

19. Yan, Q., et al.: Attention-guided deep neural network with multi-scale feature fusion for liver vessel segmentation. IEEE J. Biomed. Health. Inf. **25**(7), 2629–2642 (2021)

Hybrid Explanation-Guided Learning for Transformer-Based Chest X-Ray Diagnosis

Shelley Zixin Shu[1(✉)], Haozhe Luo[1,5], Alexander Poellinger[2,3], and Mauricio Reyes[1,4]

[1] ARTORG Center for Biomedical Engineering Research, University of Bern, Murtenstrasse 50, 3008 Bern, Switzerland
`zixin.shu@unibe.ch`
[2] Inselspital (Bern University Hospital), 3010 Bern, Switzerland
[3] Insel Gruppe Bern Universitätsinstitut für Diagnostische, Interventionelle und Pädiatrische Radiologie, Bern, Switzerland
[4] Department of Radiation Oncology, Inselspital, Bern University Hospital and University of Bern, Bern, Switzerland
[5] Kaiko.AI, Zurich, Switzerland

Abstract. Transformer-based deep learning models have demonstrated exceptional performance in medical imaging by leveraging attention mechanisms for feature representation and interpretability. However, these models are prone to learning spurious correlations, leading to biases and limited generalization. While human-AI attention alignment can mitigate these issues, it often depends on costly manual supervision. In this work, we propose a Hybrid Explanation-Guided Learning (H-EGL) framework that combines self-supervised and human-guided constraints to enhance attention alignment and improve generalization. The self-supervised component of H-EGL leverages class-distinctive attention without relying on restrictive priors, promoting robustness and flexibility. We validate our approach on chest X-ray classification using the Vision Transformer (ViT), where H-EGL outperforms two state-of-the-art Explanation-Guided Learning (EGL) methods, demonstrating superior classification accuracy and generalization capability. Additionally, it produces attention maps that are better aligned with human expertise.

Keywords: Chest X-ray Classification · Self-supervised Learning · Interpretability · Human-AI Alignment

1 Introduction

Deep learning models, particularly transformer-based architectures, have demonstrated remarkable success across various medical image computing applications. One key enabler of this success is the attention mechanism [3], which allows models to focus on the most relevant regions of an image, enabling superior feature

© The Author(s), under exclusive license to Springer Nature Switzerland AG 2026
M. Reyes et al. (Eds.): iMIMIC 2025, LNCS 16464, pp. 33–42, 2026.
https://doi.org/10.1007/978-3-032-17611-0_4

representation and interpretability. However, despite their effectiveness, deep neural networks (DNNs) are inherently data-driven and prone to learn spurious correlations, leading to shortcut learning, biases, and fairness issues [2,5,6].

Human-AI alignment integrates human knowledge into model training to align attention with human-understood features, improving robustness and generalization [14,16,18]. This can involve expert-annotated explanations or iterative human feedback. However, the high cost of human annotations remains a major challenge. Human-AI attention alignment falls under Explanation-Guided Learning (EGL) [4], which uses explanations to guide learning. To reduce annotation needs, self-supervised EGL leverages intrinsic model constraints, but risks reinforcing spurious or misaligned explanations. Contrastive self-supervised explanation learning, such as [13], lacks standardized methods for generating positive and negative samples and faces challenges in constructing such samples without ground truth [4]. These methods often rely on rigid priors like sparsity, smoothness, and stability, which, while improving interpretability, may suppress subtle or complex feature, potentially missing critical clinical cues [15].

In response to these challenges, we propose a novel Hybrid Explanation-Guided Learning (H-EGL) approach that integrates both self-supervision and human supervision for attention alignment within the Explanation-Guided Learning (EGL) paradigm. This hybrid framework facilitates human-AI alignment while leveraging unlabeled human attention data, allowing self-supervision and human guidance to complement each other effectively. A key component of H-EGL is a self-supervised EGL method called Discriminative Attention Learning (DAL), which leverages class-distinctive attention maps from Vision Transformer (ViT) models. DAL was designed with inspiration from [12]. Unlike prior self-supervised methods that impose rigid constraints, DAL introduces a flexible inductive bias that avoids over-regularization and preserves the model's ability to learn complex, task-specific features. It promotes the distinctiveness of class-specific attention maps by guiding the model to generate discriminative attention outputs, leading to more robust and generalizable representations. Rather than relying on post-hoc interpretability tools, as in [12] for convolutional networks, DAL directly exploits the inherent attention mechanisms of ViTs. To the best of our knowledge, this is the first application of a hybrid explanation-guided method within transformer architectures to assess and enhance human-AI alignment.

We evaluated our proposed method in the context of disease classification from chest X-ray images, a widely benchmarked medical imaging task. Using Vision Transformer (ViT) models, we compare our approach against two state-of-the-art EGL methods. To the best of our knowledge, this study represents the first evaluation of EGL on hybrid attention alignment for ViT models in medical imaging, which achieves superior classification performance and generates better expert-aligned attention maps. This emphasizes the complementary strengths of both paradigms.

2   Methodology

This section introduces Hybrid Explanation-Guided Learning (H-EGL), an approach that combines self-supervised and supervised EGL methods for Vision Transformer (ViT) models, applicable to a wide range of classification tasks. H-EGL consists of two key components: a self-supervised module called Discriminative Attention Learning (DAL), and a supervised module focused on human alignment. This paper demonstrate its capability in multi-label thoracic disease classification but notes that the approach is generic and can be extended to other medical image computing problems. We first present the overall framework, outlining its key components and their interactions. Then, we describe the methodology in detail, including the explanation guided learning strategy and attention alignment mechanism.

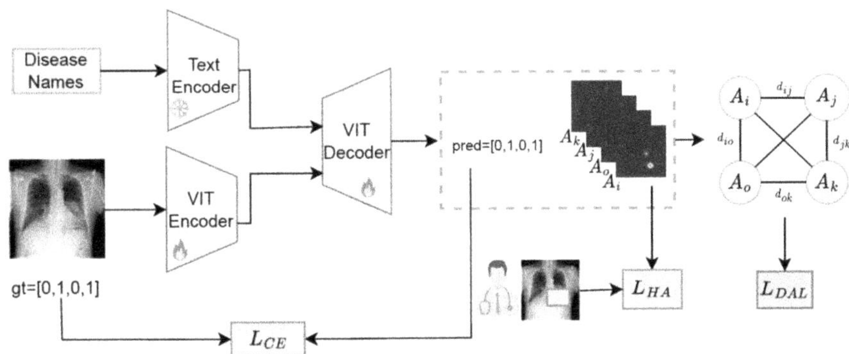

Fig. 1. Overview of the Hybrid Explanation-Guided Learning (H-EGL) framework. H-EGL is a hybrid Explanation-Guided Learning (EGL) approach designed for Vision Transformers (ViTs). The model integrates visual inputs (e.g., chest X-rays) and textual information (e.g., disease labels such as *Atelectasis, Cardiomegaly*) via a ViT-based encoder-decoder architecture, employing a frozen text encoder and trainable ViT components. It generates both class-wise predictions and attention maps A. The H-EGL framework combines two key components: (i) a self-supervised module, **Discriminative Attention Learning (DAL)**, which encourages class-distinctive attention patterns, and (ii) a supervised **Human-AI Alignment** module. The total training objective incorporates: (i) **Cross-Entropy Loss** L_{CE} for classification, (ii) **Human-AI Alignment Loss** L_{HA} to align attention maps with expert annotations (e.g., bounding boxes), and (iii) **Discriminative Attention Loss** L_{DAL}, which enforces separation among class-specific attention maps. The figure illustrates attention distinctiveness (e.g., similarity s_{ij} between attention maps A_i and A_j) in a four-class setting, though the approach generalizes to any number of classes.

We propose a unified framework, illustrated in Fig. 1, called Hybrid Explanati-on-Guided Learning (H-EGL), a novel method that integrates both human-supervised and self-supervised explanation-guided learning (EGL) for

Vision Transformer (ViT) models. It is built upon DWARF [11], a ViT-based encoder-decoder architecture tailored for disease classification from chest X-rays. DWARF fuses visual inputs (e.g., chest X-rays) and textual inputs (e.g., disease names like *Atelectasis, Cardiomegaly*) via cross-attention and aligns the model's attention maps with human-provided annotations. In this architecture, the text encoder is frozen, while the ViT encoder and decoder are trainable, allowing dynamic refinement of visual features while preserving semantic consistency from the textual side. The model outputs multi-label disease classification predictions and class-specific attention maps denoted as A.

H-EGL is designed to improve model interpretability and classification performance by jointly leveraging expert annotations and unlabeled data. It combines two complementary components: (i) a **human-AI alignment** loss that encourages the model to focus on expert-defined regions of interest, and (ii) a self-supervised module called **Discriminative Attention Learning (DAL)**, which promotes class-distinctive attention behavior across categories. The H-EGL framework is optimized using a composite loss function:

$$\mathcal{L}_{\text{H-EGL}} = \mathcal{L}_{\text{CE}} + \alpha\mathcal{L}_{\text{HA}} + \beta\mathcal{L}_{\text{DAL}}. \tag{1}$$

The weights $\alpha > 0$ and $\beta > 0$ control the relative contribution of the two explanation components. This formulation generalizes to both labeled and unlabeled human attention data and can be easily extended to other transformer-based classification tasks. $\mathcal{L}_{\text{CE}}$ is the standard multi-label cross-entropy loss for classification. $\mathcal{L}_{\text{HA}}$ is the human-AI alignment loss, aligns attention maps with expert-annotated pathology regions, adopted a penalized Dice loss following [11]:

$$\mathcal{L}_{\text{HA}} = 1 - \frac{2 \times |A_i \odot M_i|}{|A_i| + |M_i| + w_{FP}N_{FP}}, \tag{2}$$

where A_i is the model-generated attention map for class i, M_i is the corresponding expert mask, $\odot$ denotes pixel-wise multiplication, N_{FP} is the number of false positives, and w_{FP} is a penalty coefficient. $\mathcal{L}_{\text{DAL}}$ is the proposed discriminative attention learning loss, enforces class-level separability in the attention maps by minimizing their pairwise similarity,

$$\mathcal{L}_{\text{DAL}} = \frac{2}{C(C-1)} \sum_{i=1}^{C-1} \sum_{j=i+1}^{C} |S(A_i, A_j)|, \tag{3}$$

where C is the number of classes, and $S(A_i, A_j)$ denotes the cosine similarity between attention maps A_i and A_j:

$$S(A_i, A_j) = \frac{\langle \mathbf{A}_i, \mathbf{A}_j \rangle}{\|\mathbf{A}_i\| \, \|\mathbf{A}_j\|}. \tag{4}$$

This self-supervised mechanism encourages the model to produce attention maps that are more discriminative across different classes. Unlike traditional contrastive methods, DAL avoids the need for negative sample generation or image perturbation, making it efficient and scalable.

In summary, H-EGL integrates human-supervised and self-supervised attention alignment in a unified framework, improving both interpretability and performance. It supports learning from expert-labeled data while also leveraging large-scale unlabeled datasets through a lightweight, task-relevant inductive bias. This design offers a flexible and generalizable pathway toward robust human-AI aligned vision models.

3 Experiments

We evaluated H-EGL on the task of classifying four prevalent thoracic pathologies, atelectasis, cardiomegaly, consolidation, and effusion, using chest X-ray images. These conditions are well-represented in public datasets providing human-annotated attention maps, making them ideal for benchmarking [11]. We evaluate the model performance with various classification metrics and also did an ablation study by removing the individual component of the hybrid method to evaluation the individual contribution of DAL and human-AI alignment. Additionally, to evaluate model robustness and generalization, we measured the generalization gap, defined as the difference between the performance on validation and test sets [1]. We applied Gaussian noise with mean zero and various standard deviations to further examine the robustness of the model again noise at test time.

Dataset: We utilized the ChestXDet dataset [10], a subset of NIH ChestX-ray14 [17], including human-annotated pathology location segmentation. ChestXDet consists of 3,578 patients, with 3,025 samples in the training set and 553 in the test set. This official train-test split was used in our experiments. Each image includes pathology annotations (marked using bounding boxes and polygons), verified by three radiologists. We used an 80–20 train-validation split. All evaluations are conducted on the test set.

Implementation: We conducted experiments by training the model five times, each with a different random seed used to split the training and validation sets from the official training data. The models were then evaluated on the separate official test set. The reported results represent the average performance across these five runs on the official test set. We use Med-KEBERT as text encoder and Transformer Query Network decoder for cross-attention (both as part of KAD [19]). A ViT-B [3] with a 224×224 input resolution is used as an image encoder. The models achieving the highest AUC on the validation set were used for evaluation. The attention maps are obtained from the decoder's cross-attention layers. Optimization was performed using the AdamW optimizer with a learning rate of 1e-5. Models were trained for 1000 epochs with early stopping, patience set to 50 and a 20-epoch warm-up phase. The training used a batch size of 32. The penalty dice score weight w_{FP} is 1, and hyperparameters α and β were set to 1.0 in all experiments when the $\mathcal{L}_{HA}$ or $\mathcal{L}_{DAL}$ module was added for H-EGL. All experiments run on RTX 4090 GPUs with CUDA v12.2.

Baselines: We evaluated two strong baseline models in our experiments. KAD [19] is a knowledge-aware detection framework that enhances visual reasoning by incorporating human-based knowledge graphs and structured knowledge (e.g., RadGraph [8]) to better capture complex visual relationships. It serves as the foundational backbone of our model. GAIN [9] is a gradient-based attention network aimed at improving interpretable visual recognition through refined attention mechanisms. For our baseline, we adapted GAIN's loss function to work with a transformer backbone, applying its gradient-based attention strategy for EGL. We compare the performance of these baselines with our Hybrid Explanation-Guided Learning (H-EGL) method, which integrates self-supervision and human annotations for EGL. Additionally, we assess the standalone performance of DAL as a purely self-supervised EGL approach ($\alpha = 0$), and DWARF [11] ($\beta = 0$), which relies solely on human-annotation-guided explanation learning.

4   Results

We evaluated model performance using classification metrics including AUC, F_1, and MCC. To assess generalization ability, we also measured the performance gap between validation and test sets, following the methodology in [1].

Table 1 presents a comparative evaluation of H-EGL against other state-of-the-art methods. The proposed H-EGL achieved the highest test AUC (89.3%), outperforming KAD (88.1%) and the prior method GAIN (88.0%), while also exhibiting a low variance at 0.7%. Notably, H-EGL further improved classification performance, achieving the highest $F1_{\mathbf{test}}$ (69.4%) and MCC_{test} (58.3%), while significantly reducing $F1_{\mathrm{gap}}$ (0.5%) and MCC_{gap} (3.8%). These results indicate enhanced robustness and consistency across test samples.

Moreover, H-EGL outperformed both purely self-supervised (AUC 89.3±1.0%) and human-annotation-only methods (AUC 88.4%±0.2%) for explanation-guided learning (EGL). The self-supervised EGL ($\alpha = 0$) also demonstrated competitive performance relative to DWARF ($\beta = 0$), despite relying solely on self-supervised signals. These findings underscore that *H-EGL* achieves the best trade-off between performance and stability—reinforcing the effectiveness of combining self-supervised and human-guided attention constraints.

Figure 2 shows the results of a robustness analysis of models when confronted with different levels of noise added to test images. Overall, while all models show a decrease in performance at increased levels of perturbations, the proposed H-EGL approaches remain superior to the other benchmarked approaches.

Qualitative Attention Results: In addition to quantitative results, we visualized the attention maps of the baseline KAD model, DWARF, the proposed self-supervised approach (DAL), and the proposed Hybrid EGL approach (H-EGL). As shown in Fig. 3, the baseline KAD model effectively coincides with the human annotation (white dotted outline) but also wrongly highlights the lower lobe of both lungs. The DWARF model reduces false positives in the lower part of the image (see yellow arrow) but also wrongly focuses on the left lung.

Table 1. Performance and ablation comparison of H-EGL and baseline methods. This table compares the proposed H-EGL model against baseline methods: KAD [19], which utilizes knowledge graphs for improved visual reasoning, and GAIN [9], which enhances interpretability via attention guided by cross-entropy loss. An ablation study is conducted by removing key components of H-EGL: the human-annotated explanation loss (L_{HA}) and the self-supervised EGL loss (L_{DAL}). α and β refer to Eq. 1. Results are reported as mean $\pm$ standard deviation over five runs. H-EGL achieves the best overall performance across all metrics and significantly outperforms all baselines, with a paired t-test showing $p < 0.001$ compared to the next best method.

	AUC_{test} ↑	AUC_{gap} ↓	$F1_{test}$ ↑	$F1_{gap}$ ↓	MCC_{test} ↑	MCC_{gap} ↓
KAD [19]	$88.1 \pm 0.3\%$	2.5%	$68.2 \pm 2.5\%$	1.8%	$57.5 \pm 2.3\%$	4.8%
GAIN [9]	$88.0 \pm 0.4\%$	2.7%	$67.8 \pm 2.2\%$	2.4%	$57.2 \pm 2.0\%$	5.6%
H-EGL (Ours)	$\mathbf{89.3 \pm 0.7\%}$	**1.5%**	$\mathbf{69.4 \pm 1.9\%}$	**0.5%**	$\mathbf{58.3 \pm 2.5\%}$	**3.8%**
H-EGL α=0	$\mathbf{89.3 \pm 1.0\%}$	1.4%	$67.6 \pm 1.2\%$	1.4%	$56.5 \pm 1.6\%$	5.2%
β=0 [11]	$88.4 \pm 0.2\%$	2.5%	$66.9 \pm 1.2\%$	3.2%	$56.3 \pm 1.0\%$	6.5%

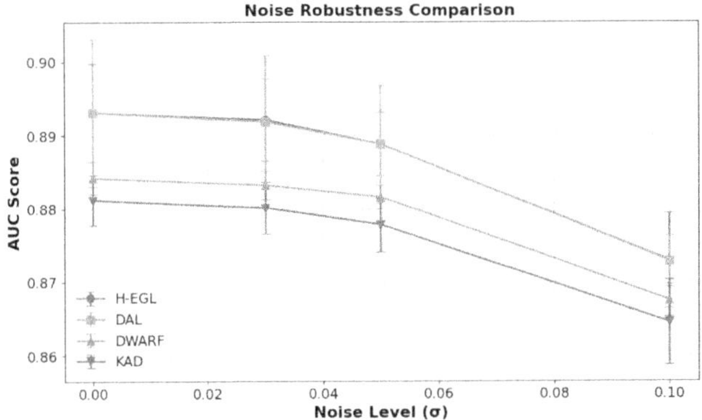

Fig. 2. Robustness of models under different levels of perturbed inputs. The noises applied are normally distributed with mean zero and different standard deviation levels ($\sigma = 0, 0.03, 0.05, 0.1$). Note: H-EGL and DAL curves closely overlap.

In contrast, H-EGL and DAL models more accurately identify the pathological regions while significantly reducing the false positives.

5   Discussion and Further Work

We introduce a Hybrid Explanation-Guided Learning (H-EGL) approach that combines self-supervision and human annotations to improve performance and generate attention maps better aligned with expert knowledge. The self-supervised component introduces an inductive bias that encourages the model to learn distinctive, class-specific attention patterns. H-EGL is architecture-agnostic and flexible. Its self-supervised component DAL imposing no constraints

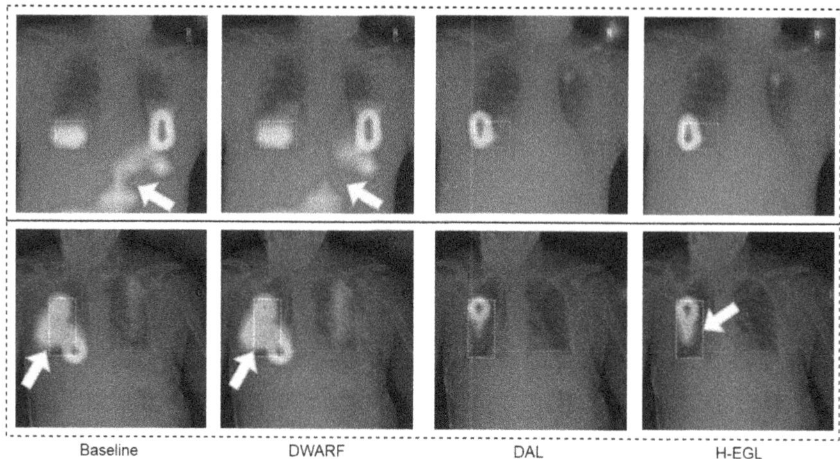

Fig. 3. Comparison of attention maps from the baseline model ($\alpha = 0$, $\beta=0$), DWARF ($\beta = 0$), DAL ($\alpha = 0$), and H-EGL model. Abnormal pathologies are highlighted with white bounding boxes, while arrows indicate areas with visible changes across models.

on feature selection, requiring no localization annotations, and eliminating the need for explicitly defined negative samples.

Our experiments show that incorporating discriminative attention guidance significantly boosts AUC scores, with performance dropping noticeably when this component is removed. H-EGL also demonstrates strong robustness under noisy test conditions, consistently outperforming baseline methods. By leveraging both labeled and unlabeled data, the framework improves attention quality and classification accuracy while remaining scalable, annotation-efficient, and compatible with existing transformer-based medical imaging models.

A key insight from our findings is the need to balance human alignment and self-supervision for explanation-guided learning. While human-guided alignment steers the model toward clinically meaningful features, fully supervised methods can be costly and may result in rigid attention behaviors that lack generalization. On the other hand, purely self-supervised strategies, though scalable, risk encouraging shortcut learning when unconstrained. H-EGL addresses this by integrating both paradigms, enabling human-guided supervision to refine model explanations without overly restricting its learning capacity. This approach parallels findings in policy learning, where Trust Region (TR) methods prevent overoptimization by adaptively adjusting constraints during training [7]. Similarly, H-EGL strikes a balance between autonomy and guidance, promoting both robustness and generalization.

We further observe that DWARF faces a trade-off between interpretability and performance: strong human alignment can reduce classification accuracy (see Table 1 and Fig. 3). In contrast, H-EGL effectively localize pathological regions without compromising accuracy, achieving a better balance between interpretability and predictive power. Inspired by policy learning strategies, future work will explore dynamic mechanisms to optimize the degree of self-supervision and human alignment during training, further enhancing interpretability, generalization, and robustness in medical image analysis.

6   Conclusion

We propose a hybrid attention alignment approach to improve the robustness and interpretability of transformer-based medical imaging models. The Hybrid Explanation-Guided Learning (H-EGL) framework combines human-guided and self-supervised learning to localize pathological regions and boost classification performance. Results validate H-EGL as a scalable alternative to fully supervised alignment. Future work will extend to larger datasets and explore dynamic alignment strategies to further enhance interpretability and generalization in clinical tasks.

Acknowledgments. We acknowledge funding by the Swiss National Science Foundation (project number 212939). We report no financial relationship or conflicts of interest.

References

1. D'Amour, A., et al.: Underspecification presents challenges for credibility in modern machine learning. J. Mach. Learn. Res. **23**(226), 1–61 (2022)
2. DeGrave, A.J., Janizek, J.D., Lee, S.I.: Ai for radiographic covid-19 detection selects shortcuts over signal. Nat. Mach. Intell. **3**(7), 610–619 (2021)
3. Dosovitskiy, A., et al.: An image is worth 16x16 words: transformers for image recognition at scale. arXiv preprint arXiv:2010.11929 (2020)
4. Gao, Y., Gu, S., Jiang, J., Hong, S.R., Yu, D., Zhao, L.: Going beyond xai: a systematic survey for explanation-guided learning. ACM Comput. Surv. **56**(7), 1–39 (2024)
5. Geirhos, R., Jacobsen, J.H., Michaelis, C., Zemel, R., Brendel, W., Bethge, M., Wichmann, F.A.: Shortcut learning in deep neural networks. Nat. Mach. Intell. **2**(11), 665–673 (2020)
6. Gichoya, J.W., et al.: Ai recognition of patient race in medical imaging: a modelling study. Lancet Digit. Health **4**(6), e406–e414 (2022)
7. Gorbatovski, A., et al.: Learn your reference model for real good alignment. arXiv preprint arXiv:2404.09656 (2024)
8. Jain, S., et al.: Radgraph: extracting clinical entities and relations from radiology reports. arXiv preprint arXiv:2106.14463 (2021)
9. Li, K., Wu, Z., Peng, K.C., Ernst, J., Fu, Y.: Tell me where to look: Guided attention inference network. In: Proceedings of the IEEE Conference on Computer Vision and Pattern Recognition, pp. 9215–9223 (2018)
10. Liu, J., Lian, J., Yu, Y.: Chestx-det10: chest x-ray dataset on detection of thoracic abnormalities. arXiv preprint arXiv:2006.10550 (2020)
11. Luo, H., de Mortanges, A.P., Inel, O., Reyes, M.: Dwarf: disease-weighted network for attention map refinement. In: International Conference on Medical Image Computing and Computer-Assisted Intervention, pp. 59–68. Springer (2024)

12. Mahapatra, D., Poellinger, A., Reyes, M.: Interpretability-guided inductive bias for deep learning based medical image. Med. Image Anal. **81**, 102551 (2022)
13. Pedapati, T., Balakrishnan, A., Shanmugam, K., Dhurandhar, A.: Learning global transparent models consistent with local contrastive explanations. Adv. Neural. Inf. Process. Syst. **33**, 3592–3602 (2020)
14. Rieger, L., Singh, C., Murdoch, W., Yu, B.: Interpretations are useful: penalizing explanations to align neural networks with prior knowledge. In: International Conference on Machine Learning, pp. 8116–8126. PMLR (2020)
15. Schramowski, P., et al.: Making deep neural networks right for the right scientific reasons by interacting with their explanations. Nat. Mach. Intell. **2**(8), 476–486 (2020)
16. Wang, S., Ouyang, X., Liu, T., Wang, Q., Shen, D.: Follow my eye: using gaze to supervise computer-aided diagnosis. IEEE Trans. Med. Imaging **41**(7), 1688–1698 (2022)
17. Wang, X., Peng, Y., Lu, L., Lu, Z., Bagheri, M., Summers, R.M.: Chestx-ray8: hospital-scale chest x-ray database and benchmarks on weakly-supervised classification and localization of common thorax diseases. In: Proceedings of the IEEE Conference on Computer Vision and Pattern Recognition, pp. 2097–2106 (2017)
18. Wu, S., Zhang, X., Wang, B., Jin, Z., Li, H., Feng, J.: Gaze-directed vision gnn for mitigating shortcut learning in medical image. In: International Conference on Medical Image Computing and Computer-Assisted Intervention, pp. 514–524. Springer (2024)
19. Zhang, X., Wu, C., Zhang, Y., Xie, W., Wang, Y.: Knowledge-enhanced visual-language pre-training on chest radiology images. Nat. Commun. **14**(1), 4542 (2023)

Interpretable Networks to Model the Accumulation of Tau Protein in the Brain

Nicolas Honnorat and Mohamad Habes[(✉)]

Department of Radiology, University of Texas Health Science Center at San Antonio, San Antonio, USA
habes@uthscsa.edu

Abstract. Recent advances in tau PET imaging offer new insight into the accumulation of neurofibrillary tangles in the brain associated with dementia. Large datasets are now available to create advanced statistical models, such as Deep Learning models. Unfortunately, most of the Deep Networks derived so far from tau PET data were designed to achieve complex tasks, such as classification and regression tasks, that require convoluted analysis to produce interpretable brain maps. In this work, we propose a more straightforward approach to model the presence of tau protein in the brain: we use small interpretable Deep Networks to model the spatial variation across the cortex of probability distributions derived from a large population of 1,386 ADNI tau PET scans. We compare 192 variants of network architectures with standard statistical approaches and select a reliable model to generate brain maps describing how pathological tau proteins accumulate in the cortex.

Keywords: PET · tau · neurofibrillary tangles · Deep Learning

1 Introduction

Alzheimer's disease (AD) is characterized by two types of pathological protein accumulations in the brain: amyloid beta plaques and tau neurofibrillary tangles [4,28]. Positron Emission Tomography (PET) tracers sensitive to tau proteins, such as flortaucipir [5], were just developed a few years ago, but large clinical studies, such as the Alzheimer's Disease Neuroimaging Initiative (ADNI [1]), have massively invested in these new tracers. This recent profusion of data has offered the possibility of using advanced statistical models to describe tau accumulation [13] and tau propagation in the brain [22,27]. Deep Learning plays a crucial role in many such statistical analyses [7,17,24,33]. Deep Networks are typically used to create efficient models performing advanced tasks, such as diagnosing Alzheimer's disease [7], synthesizing tau PET scans from other scans [17], or matching tau and amyloid deposits [24]. Unfortunately, these models require an additional and time-consuming effort to interpret, such as the derivation of heatmaps indicating the variables selected by classification models [33].

© The Author(s), under exclusive license to Springer Nature Switzerland AG 2026
M. Reyes et al. (Eds.): iMIMIC 2025, LNCS 16464, pp. 43–52, 2026.
https://doi.org/10.1007/978-3-032-17611-0_5

In this work, we propose a different approach: starting from a set of PET signal distributions capturing the accumulation of tau protein in thousand brain locations that are accurate but difficult to interpret because of their high dimensionality, we use Deep Networks to produce compact and interpretable representations while maintaining a good accuracy during the approximation. More specifically, our networks are trained to predict the quantiles of the PET signal distributions from a set of signed geodesic distances that we use to encode spatial location across the cortex. The networks are designed to produce a set of attention maps [32] revealing cortical regions associated with distinct tau accumulation patterns. Our experiments demonstrate that our interpretable networks produce stable results and achieve superior reconstruction accuracies compared to standard machine learning approaches.

2   Materials and Methods

2.1   Tau PET Processing

1,386 tau PET scans provided by the Alzheimer's Disease Neuroimaging Initiative (ADNI) were collected [14]. They correspond to a population with a mean age of 74.03 years (standard deviation 7.62 years) and 52.6% women.

For each PET scan, the T1-weighted brain MRI scan acquired within the shortest time period was found and used to conduct the following preprocessing. T1-weighted scans were first bias field corrected using the N4 method [29] distributed within the Advanced Normalization Tools software (ANTS [3], version 2.2.0) library. Non-local mean filtering was conducted using NAONLM3D [18], and the HD-BET neural network was used in fast CPU mode to segment the brains ("-device cpu -mode fast -tta 0" command parameters [12]). The SyN registration method [2] provided with ANTS was then used to compute non-rigid deformation fields warping these individual brains into the space of the Conte69 atlas [31] where the Human Connectome Project (HCP) has delineated cerebellar regions and defined two cortical meshes and their parcellations [8,30].

The tau PET scans were then processed as follows. ANTS was used to find a rigid transformation that aligns PET scans in the space of their associated T1-weighted scans [3]. HD-BET brain masks were used to mask the PET signal outside the brain [12]. The non-rigid transforms were then applied to warp the masked PET scans within the space of the Conte69 atlas [31], where the Connectome Workbench (version 1.5.0) was used to project the PET signals onto the cerebellum grey matter regions [8] and the two HCP meshes, a mesh capturing the 3D surface of the cortex in the left brain hemisphere and a mesh describing the right cortical surface [19]. The projected data were then scaled separately for each scan to account for variations in PET tracer dose. This was achieved by concatenating the PET signals measured in the cerebellar grey matter regions of the two hemispheres, calculating the median of these measurements, and dividing the cortical PET signals by this median value [13,22].

2.2   Quantiles and Geodesic Maps

To model the accumulation of tau protein over time, the processed PET scans were divided into five even age groups: 4 groups with 277 scans and 1 group with 278 scans. In each group, the scans acquired for men and women were separated, and 250 scans were randomly selected with replacement for each sex. This bootstrapping generated five groups of five hundred scans balanced to mitigate sex biases so that our models would focus on aging and AD effects.

A PET signal distribution was modeled for each age group and each cortical location defined in the HCP mesh of each hemisphere [8] by collecting the PET signal values of the bootstrapped age group scans. These five hundred values were sorted in increasing order, and twenty quantile values were computed. Concatenating the quantiles obtained for the five age groups produced a single vector of a hundred values for each cortical location, capturing the distribution of tau tangles in that location across the ADNI population and across time/age.

The spatial location of the cortical mesh nodes was encoded by calculating twenty-two cortical maps for each hemisphere, storing the signed geodesic distance to the twenty-two large brain regions defined in the HCP parcellations [8]. These maps were calculated via region-growing: for each hemisphere and each large HCP parcellation region, the cortical mesh nodes part of the region and directly connected to cortical nodes labeled in other regions were set to a geodesic distance of 0. The cortical nodes connected to this region border were then set to a geodesic distance of 1 outside the region and -1 within the region. The process was repeated until all the cortical locations were reached. Since we were only interested in cortical locations (29,696 mesh nodes on the left hemisphere, 29,716 nodes on the right hemisphere), the mesh nodes in the corpus callosum were ignored during the geodesic distance calculations. Figure 1 presents the signed geodesic maps calculated for four parcellation regions [8].

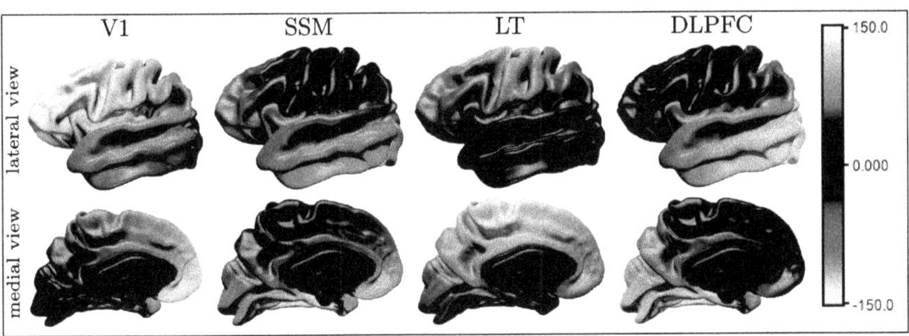

Fig. 1. Signed geodesic distance maps computed on the left hemisphere for the V1 region of the visual cortex, the somatosensory and motor region (SSM), the lateral temporal region (LT), and the dorsolateral prefrontal cortex (DLPFC) [8].

2.3   Interpretable Spatial Attention Networks

Considering each cortical location as an individual sample, we used Machine Learning to build models predicting the hundred quantile values from the twenty-two geodesic distances at that location. We also used the Deep Network architecture shown in Fig. 2 for the task. This network is made of fully connected layers with output dimensions d and leaky RELU [21] activation followed by batch normalization layers [11]. The second last fully connected layer is set with softmax activation to produce d probabilities and followed by a fully connected layer with linear activation producing the quantile estimates.

These new networks are straightforward to interpret: quantile estimates are produced by the convex combination of d weight vectors in the last fully connected layer, each vector encoding a tau distribution varying over time (accumulation pattern). This combination of temporal accumulation patterns is controlled by the probabilities generated by the first network layers. These probabilities can be interpreted as mixture parameters and vary according to the spatial location encoded in the geodesic distances. Calculating probabilities for the entire cortex produces d brain maps similar to attention maps [32] and indicating where the temporal accumulation patterns are observed in the cortex.

During our experiments, layer parameters were initialized by sampling from the $N(0, 0.05)$ distribution. An L2 kernel regularizer of amplitude λ was introduced to penalize the weights of the last fully connected layer to reduce the model variability (by forcing networks to select patterns with similar L2 norms). **A grid search was conducted to set the three parameters defining the network architecture**: the number n of fully-connected layers producing the probabilities, the number of patterns d, and the penalty λ. All networks were implemented using the Python Tensorflow library (version 2.18.0) and default leaky RELU parameters. They were optimized to minimize the mean absolute error between predicted and expected quantile values. The optimization was carried out using the Adam accelerated gradient method set to its default parameter values [15], for a batch size of 512, four learning rates of decreasing values 10^{-2}, 10^{-3}, 10^{-4}, and 10^{-5} and for 250 epochs per learning rate (1,000 epochs total).

2.4   Experiments

The performance of the networks was measured by conducting an **inpainting task** designed like a Monte Carlo cross-validation [16]: for both hemispheres, twenty-five times in a row, the cortical locations were randomly divided in a set of 80% training and 20% testing samples. The training set was used to fit the models, and a mean absolute error was measured for the testing set as follows, where t is the number of testing samples, q_i^j and p_i^j the real and predicted quantile values number j for the sample i:

$$MAE = \frac{1}{t} \sum_{i=1}^{t} \sum_{j=1}^{100} |q_i^j - p_i^j| \tag{1}$$

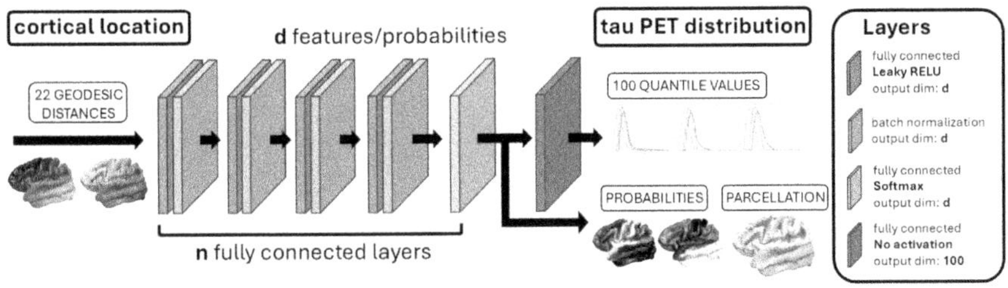

Fig. 2. Deep Network architecture adopted to model the tau protein accumulation in the cortex. Networks are made of a sequence of fully connected layers with leaky RELU activations followed by batch normalization layers. A fully connected layer with a softmax activation generates probability maps that can be summarized in a parcellation, and the last fully connected layer predicts the tau PET signal distribution quantiles.

Even though this task is not a rigorous cross-validation because signals measured in adjacent brain locations are not statistically independent, MAE indicates if the spatial patterns of tau accumulation are correctly captured.

Four numbers of layers n were tested (2,3,4,5), as well as 8 d values (4,8,12,16, 20,24,28,32) and 6 penalties λ $(0,10^{-5},10^{-4},10^{-3},10^{-2},10^{-1})$. The MAE obtained by these 192 architectures was compared with the MAE measured when the test quantile values were directly predicted as a mean or a median of the training quantile values. They were also compared with the errors measured when linear models were used to predict the quantile values from the geodesic distance maps. Two sets of linear models were tested: linear models derived via ordinary least square (OLS) and a set of models predicting each quantile value via a robust linear regression and implemented using Python statsmodels API (version 0.12.2) and using a robust norm based on Huber's estimator [9]. In addition, nonlinear Support Vector regressors relying on a Radial Basis Function kernel [6,25] and implemented using the Scikit-learn Python library [23] (version 1.6.0) were tested. Five C parameter values were tested for these SVRs: 1, 10^1, 10^2, 10^3, 10^4, and 10^5. SVRs were trained using a random tenth of the training data to save computational time.

The stability of Deep Network models was measured by applying the trained models to the entire cortical hemisphere and generating a parcellation indicating the highest probability in each mesh node. This approach produced twenty-five cortical parcellations for each hemisphere, which were compared by calculating an adjusted Rand score for each pair of parcellation [10,26]. Adjusted Rand (aR) scores measure the similarity between parcellations and take values between -1 and 1 (perfect similarity, up to a label permutation). They were calculated by computing the contingency table m_{xy} counting the number of brain locations labeled as x in the first parcellation and y in the second parcellation, and the following sums [10, 26]:

$$a_x = \sum_y m_{xy} \qquad b_y = \sum_x m_{xy} \qquad m = \sum_x a_x = \sum_y b_y \tag{2}$$

$$aR = \frac{\sum_{xy} \binom{m_{xy}}{2} - \left[\sum_x \binom{a_x}{2} \sum_y \binom{b_y}{2} \right] / \binom{m}{2}}{\frac{1}{2} \left[\sum_x \binom{a_x}{2} + \sum_y \binom{b_y}{2} \right] - \left[\sum_x \binom{a_x}{2} \sum_y \binom{b_y}{2} \right] / \binom{m}{2}} \tag{3}$$

The medians of the 300 adjusted Rand scores obtained for each hemisphere were reported (all pairs of 25 parcellations). Training time was also recorded.

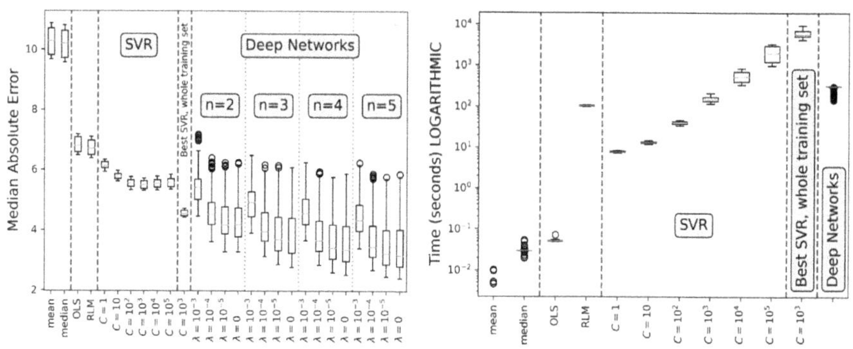

Fig. 3. Median absolute error (MAE) and training time measured for all the methods tested. To save space, MAE is only presented for the smallest four λ penalties.

3 Results

3.1 Training Time and Mean Absolute Errors

Figure 3 presents the prediction errors and the training times. Deep Networks with more parameters were slightly slower to train, but these differences were negligible compared to the training time measured for other methods. Despite random subsampling, SVRs were extremely slow to train and never reached the performance of the best Deep Networks, even when using all training samples. Deep Network prediction errors are reported in the next Sect. 3.2.

3.2 Deep Network Architecture Selection

Overall, 9600 Deep Networks corresponding to 192 different architectures were trained and tested during our inpainting experiments. Figure 4 presents the median MAE prediction errors and adjusted Rand (aR) scores obtained for all of them. Since the values obtained for the left and right hemispheres were similar, the two MAE and adjusted Rand values were averaged. The deepest networks

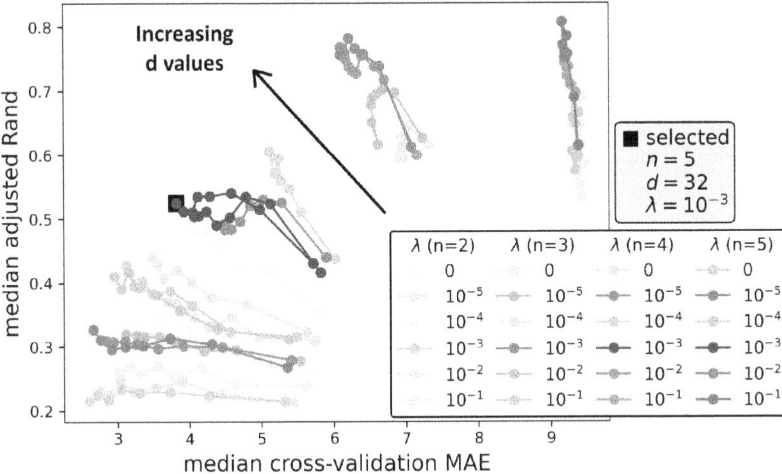

Fig. 4. Inpainting median absolute error (MAE) and stability measured by the adjusted Rand for the 192 Deep Network architectures tested in this work. The architecture marked with the black square was selected to generate the final results.

with the largest d values produced the best predictions, but Fig. 4 indicates that models using four or five layers produced similar results. Similarly, d values between 24 and 32 are associated with similar performances. Figure 4 also highlights the crucial role of the λ penalty introduced to penalize the last fully connected layer. Small penalties produce small MAE prediction errors but probability maps are randomly scaled. As a result, parcellations are very unstable. In contrast, large penalties produce models focusing on a few reliable patterns to represent the entire cortex, which yields good aR values but poor MAE. When λ increases, stability increases first. Significant MAE degradations only happen for penalties 10^{-3} and larger. This penalty was then chosen to select an architecture that was simultaneously fairly stable and accurate: the architecture with $n = 5$ layers, $d = 32$, and penalty $\lambda = 10^{-3}$ was selected, with an aR score above 0.5 and an MAE just a third larger than the best errors.

3.3 Final Results

Two networks were trained: one network using all the data available for the left hemisphere, which was applied to the entire set of geodesic distances calculated for that hemisphere, and one network for the right hemisphere applied to the right hemisphere. For visualization purposes, the 32 probability maps generated by the two networks were matched as follows. For each map, an average value was calculated in each of the 180 cortical parcels of the high-resolution HCP parcellation [8]. The inner products between the 32 sets of 180 average values obtained for the left hemisphere and the 32 sets of 180 values of the right hemisphere were stored in a 32×32 similarity matrix. This matrix was converted

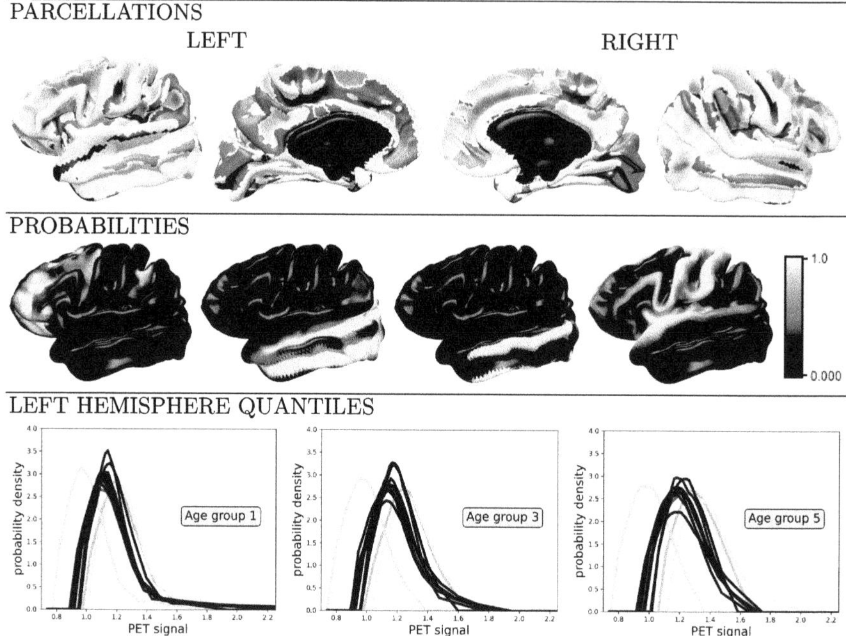

Fig. 5. For the selected architecture: parcellations obtained for both hemispheres, probability maps of four salient regions (orange/frontal, yellow/temporal+precuneus, brown/medial temporal gyri, and grey/sensorymotor+superior temporal gyri), and tau distributions across the youngest, middle-aged, and elder age groups. (Color figure online)

into a cost matrix by subtracting these values from their maximum, and the Hungarian algorithm was used to find the matching minimizing the costs [20].

Final results are shown in Fig. 5. Brain maps indicate different tau accumulations in gyri and sulci, in temporal, medial, and frontal regions. For instance, the grey region covering the somatosensory, superior temporal, and precentral gyri is associated with a very small and stable presence of tau proteins. In contrast, temporal sulci and the precuneus area (in yellow) exhibit strong tau PET signals and a strong increase over time. A few ADNI participants also present accelerated deposition in these areas, as indicated by the fast shift in time of the top quantiles toward large values. These results are coherent with prior literature [4,13,22]. Overall, by selecting thin gyri and deep sulci, our Deep Networks demonstrate their ability to localize precise brain regions in the cortex. Their implementation will be shared via our lab GitHub account (https://github.com/UTHSCSA-NAL/imimic2025).

4 Conclusion

In this work, we used Deep Networks to generate interpretable brain maps and temporal patterns capturing the accumulation of pathological tau proteins in the brain. These models revealed distinct accumulation patterns and identified cortical areas exhibiting accelerated deposits. Our results indicate that our networks produce more accurate spatial models than basic machine learning methods. Future investigations will be conducted to establish if the patterns of accelerated tau deposits revealed in this work are associated with severe cognitive decline.

References

1. https://adni.loni.usc.edu/data-samples/adni-data/neuroimaging/pet/
2. Avants, B., Epstein, C., Grossman, M., Gee, J.: Symmetric diffeomorphic image registration with cross-correlation: evaluating automated labeling of elderly and neurodegenerative brain. Med. Image Anal. **12**(1), 26–41 (2008)
3. Avants, B., Tustison, N., Wu, J., Cook, P., Gee, J.: An open source multivariate framework for n-tissue segmentation with evaluation on public data. Neuroinformatics **9**(4), 381–400 (2011)
4. Braak, H., Braak, E.: Neuropathological stageing of Alzheimer-related changes. Acta Neuropathol. **82**(4), 239–259 (1991)
5. Burnham, S.C., et al.: A review of the flortaucipir literature for positron emission tomography imaging of tau neurofibrillary tangles. Brain Commun. **6**(1), fcad305 (2023)
6. Drucker, H., Burges, C.J., Kaufman, L., Smola, A., Vapnik, V.: Support vector regression machines. Adv. Neural Inform. Process. Syst. **9** (1996)
7. Elaza, A., et al.: Alzheimer's disease diagnosis from single and multimodal data using machine and deep learning models: Achievements and future directions. Expert Syst. Appl. **255**, 124780 (2024)
8. Glasser, M., et al.: A multi-modal parcellation of human cerebral cortex. Nature **536**, 171–178 (2016)
9. Huber, P.J.: Robust estimation of a location parameter. Ann. Math. Stat. **35**, 73–101 (1964)
10. Hubert, L., Arabie, P.: Comparing partitions. J. Classif. **2**(1), 193–218 (1985)
11. Ioffe, S., Szegedy, C.: Batch normalization: accelerating deep network training by reducing internal covariate shift. In: ICML, pp. 448 – 456 (2015)
12. Isensee, F., et al.: Automated brain extraction of multisequence mri using artificial neural networks. Hum. Brain Mapp. **40**(17), 4952–4964 (2019)
13. Jack, C.J., et al.: Longitudinal tau pet in ageing and alzheimer's disease. Brain **141**(5), 1517–1528 (2018)
14. Jagust, W., et al.: The Alzheimer's Disease neuroimaging initiative 2 PET core: 2015. Alzheimer's & DementiaâĂŕ: J, Alzheimer's Association **11**(7), 757–771 (2015)
15. Kingma, D.P., Ba, J.: Adam: a method for stochastic optimization. In: Proceedings of the 3rd International Conference on Learning Representations (2015)
16. Kuhn, M., Johnson, K.: Applied Predictive Modeling. Springer New York (2013)
17. Lee, J., et al.: Synthesizing images of tau pathology from cross-modal neuroimaging using deep learning. Brain **147**, 980–995 (2024)

18. Manjón, J., Coupé, P., Martí-Bonmatí, L., Collins, D., Robles, M.: Adaptive non-local means denoising of mr images with spatially varying noise levels. J. Magnetic Resonance Imaging: JMRI **31**(1), 192–203 (2010)
19. Marcus, D., et al.: Informatics and data mining tools and strategies for the human connectome project. Front. Neuroinform. **5**, 4 (2011)
20. Munkres, J.: Algorithms for the assignment and transportation problems. J. Soc. Ind. Appl. Math. **5**(1), 32–38 (1957)
21. Nair, V., Hinton, G.: Rectified linear units improve restricted Boltzmann machines. In: ICML (2010)
22. Ottoy, J., et al.: Tau follows principal axes of functional and structural brain organization in alzheimer's disease. Nat. Commun. **15**, 5031 (2024)
23. Pedregosa, F., et al.: Scikit-learn: machine learning in Python. J. Mach. Learn. Res. **12**, 2825–2830 (2011)
24. Ruwanpathirana, G., Williams, R., Masters, C., Rowe, C., Johnston, L., Davey, C.: Mapping the association between tau-pet and aβ-amyloid-pet using deep learning. Sci. Rep. **12**(1), 14797 (2022)
25. Smola, A., Schölkopf, B.: A tutorial on support vector regression. Statist. Comput. Archive **14**(3), 199–222 (2004)
26. Steinley, D.: Properties of the Hubert-Arabie adjusted rand index **9**(3), 386–396 (2004)
27. Steward, A., et al.: Functional network segregation is associated with attenuated tau spreading in alzheimer's disease. Alzheimer's & DementiaâĂŕ: J. Alzheimer's Associat. **19**(5), 2034–2046 (2023)
28. Thal, D., Rüb, U., Orantes, M., Braak, H.: Phases of a beta-deposition in the human brain and its relevance for the development of AD. Neurology **58**(12), 1791–1800 (2002)
29. Tustison, N.: N4ITK: improved n3 bias correction. IEEE Trans. Med. Imaging **29**(6), 1310–1320 (2010)
30. Van Essen, D.C., Smith, S., Barch, D., Behrens, T., Yacoub, E., Ugurbil, K.: The WU-Minn human connectome project: An overview. Neuroimage **80**, 62–79 (2013)
31. Van Essen, D., Glasser, M., Dierker, D., Harwell, J., Coalson, T.: Parcellations and hemispheric asymmetries of human cerebral cortex analyzed on surface-based atlases. Cerebral Cortex **22**(10), 2241 – 2262 (2012)
32. Vaswani, A., et al.: Attention is all you need. In: Guyon, I., Luxburg, U.V., Bengio, S., Wallach, H., Fergus, R., Vishwanathan, S., Garnett, R. (eds.) Advances in Neural Information Processing Systems. vol. 30. Curran Associates, Inc. (2017)
33. Wang, D., et al.: Deep learning reveals pathology-confirmed neuroimaging signatures in Alzheimer's, vascular and Lewy body dementias. Brain p. awae388 (2024)

Segmenting Thalamic Nuclei: T_1 Maps Provide a Reliable and Efficient Solution

Anqi Feng[1,2], Zhangxing Bian[1], Samuel W. Remedios[3], Savannah P. Hays[1], Blake E. Dewey[4], Jiachen Zhuo[5], Dan Benjamini[2], and Jerry L. Prince[1(✉)]

[1] Department of Electrical and Computer Engineering, Johns Hopkins University, Baltimore, USA
{afeng11,prince}@jhu.edu
[2] Laboratory of Behavioral Neuroscience, National Institute on Aging, National Institutes of Health, Baltimore, USA
[3] Department of Computer Science, Johns Hopkins University, Baltimore, USA
[4] Department of Neurology, Johns Hopkins School of Medicine, Baltimore, USA
[5] Department of Diagnostic Radiology and Nuclear Medicine, University of Maryland School of Medicine, Baltimore, USA

Abstract. Accurate thalamic nuclei segmentation is crucial for understanding neurological diseases, brain functions, and guiding clinical interventions. However, the optimal inputs for segmentation remain unclear. This study systematically evaluates multiple MRI contrasts, including MPRAGE and FGATIR sequences, quantitative PD and T_1 maps, and multiple T1-weighted images at different inversion times (multi-TI), to determine the most effective inputs. For multi-TI images, we employ a gradient-based saliency analysis with Monte Carlo dropout and propose an Overall Importance Score to select the images contributing most to segmentation. A 3D U-Net is trained on each of these configurations. Results show that T_1 maps alone achieve strong quantitative performance and superior qualitative outcomes, while PD maps offer no added value. These findings underscore the value of T_1 maps as a reliable and efficient input among the evaluated options, providing valuable guidance for optimizing imaging protocols when thalamic structures are of clinical or research interest. Codes are available at T1map-for-Thalamus.

Keywords: Thalamic Nuclei Segmentation · MRI Contrast Selection · Gradient-Based Saliency Analysis · Monte Carlo Dropout

1 Introduction

The thalamus is a deep brain structure consisting of multiple distinct nuclei, each specializing in functions and involved in neurological disorders and treatments [27,28]. For example, the ventral lateral posterior nucleus regulates motor control [24,29]; the limbic nuclei relate to Alzheimer's disease [1,3]; and the ventral intermediate nucleus is a key target for deep brain simulation (DBS) in essential tremor [16,23]. Given these roles, accurate segmentation of thalamic nuclei is important for both clinical interventions and neurological research.

© The Author(s), under exclusive license to Springer Nature Switzerland AG 2026
M. Reyes et al. (Eds.): iMIMIC 2025, LNCS 16464, pp. 53–63, 2026.
https://doi.org/10.1007/978-3-032-17611-0_6

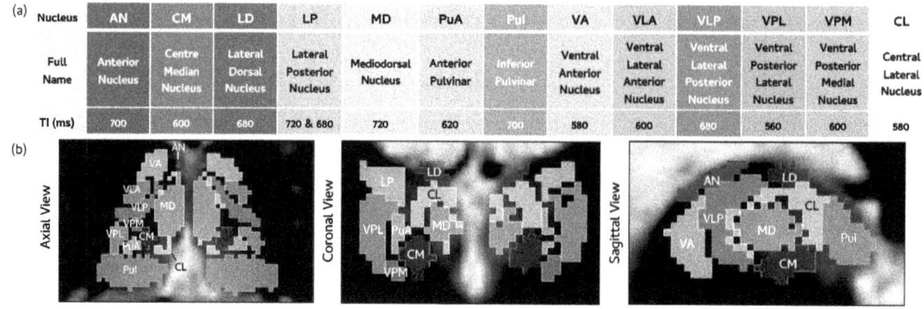

Nucleus	AN	CM	LD	LP	MD	PuA	Pul	VA	VLA	VLP	VPL	VPM	CL
Full Name	Anterior Nucleus	Centre Median Nucleus	Lateral Dorsal Nucleus	Lateral Posterior Nucleus	Mediodorsal Nucleus	Anterior Pulvinar	Inferior Pulvinar	Ventral Anterior Nucleus	Ventral Lateral Anterior Nucleus	Ventral Lateral Posterior Nucleus	Ventral Posterior Lateral Nucleus	Ventral Posterior Medial Nucleus	Central Lateral Nucleus
TI (ms)	700	600	680	720 & 680	720	620	700	580	600	680	560	600	580

Fig. 1. (a) Thalamic nuclei with their color codes, and corresponding inversion times (TI) used for manual delineation. (b) Sparse labels overlaid on the T_1 map.

In-vivo thalamic nuclei segmentation primarily relies on Magnetic Resonance Imaging (MRI). However, low intra-thalamic contrast and small nuclei size make accurate segmentation challenging with standard T1-weighted (T1w) and T2-weighted (T2w) sequences. Magnetization-Prepared Rapid Gradient Echo (MPRAGE) [19] has been widely used but provides limited contrast within the thalamus and against surrounding white matter and subcortical structures [13,35]. Fast Gray Matter Acquisition T1 Inversion Recovery (FGATIR) [31] has been shown to enhance intra-thalamic contrast by nulling white matter [25,30,33], though the shorter inversion time can lead to a somewhat lower signal-to-noise (SNR) ratio compared to MPRAGE [33]. Quantitative MRI, which can yield intrinsic tissue properties such as T_1, T_2, and proton density (PD) maps, has also been explored [6,34]. T_1 maps, in particular, reveal relaxation differences between nuclei [5,33], facilitating visual discrimination of major nuclear groups. Some methods acquire multiple T1-weighted images at different inversion times (multi-TI), each tailored to null specific tissue compartments at different timepoints, capturing finer contrast variations among nuclei [9,10,37]. Diffusion MRI (dMRI) has been widely used in [7,21,36], but the low anisotropy in thalamic gray matter and the limited spatial resolution of Echo Planar Imaging (EPI) hinder accurate nuclei localization. Connectivity-based segmentation is clinically useful for DBS targeting [14,22], but often misaligns with anatomical boundaries. Many adjacent nuclei project to similar cortical targets, grouping distinct nuclei together, while individual nuclei with diverse projection patterns may split into multiple clusters.

Multi-TI images offer diverse tissue contrasts across inversion times. However, not all are equally informative for thalamic nuclei segmentation. Selecting the most relevant images is necessary, but training a separate model for every possible combination is impractical. Gradient-based methods are widely used to generate post-hoc attribution maps for identifying important image regions [26,32], and have been extended to assess the importance of multimodal medical images [15]. Monte Carlo Dropout [11] estimates prediction uncertainty at test time through multiple stochastic forward passes and has been applied in medical imaging tasks [8,20]. However, integrating gradient-based saliency analysis with Monte Carlo Dropout to quantify image importance remains underexplored.

Despite previous efforts, the optimal MRI contrast for thalamic nuclei segmentation is still unclear. In this study, we focus on T1w-based images, systematically comparing structural sequences (MPRAGE, FGATIR), quantitative maps (PD, T_1 maps), and multi-TI images. For multi-TI images, we perform gradient-based saliency analysis across multiple Monte Carlo dropout runs and define an Overall Importance Score to select the most informative inversion times for segmentation. To the best of our knowledge, this is the first study to leverage gradient-based saliency analysis with Monte Carlo Dropout to guide input feature selection. Our results show that the T_1 maps alone strike a favorable balance between accuracy and efficiency, making it a practical choice for thalamic nuclei segmentation.

2   Methods

2.1   Data and Manual Labels

Our MRI data were collected as part of a mild traumatic brain injury (MTBI) study, approved by the local ethics board. It involves 24 participants: 14 healthy controls and 10 individuals with MTBI. Each participant underwent an MRI session with MPRAGE and FGATIR, both acquired at 1 mm isotropic resolution, repetition time (TR) 4000 ms, and echo time (TE) 3.37 ms. The only difference was in the inversion time (TI): 1400 ms for MPRAGE and 400 ms for FGATIR.

Manual thalamic nuclei delineation was guided by the Morel Atlas [18], focusing on 13 nuclei (or nuclear groups). Multi-TI images were used for annotation due to their enhanced contrast for these small, densely packed structures. For each nucleus, the rater selected the most visually distinct TI sequence to maximize delineation clarity, then verified alignment with T_1 maps. Figure 1(a) lists the nuclei names along with the chosen TI values. Only voxels with high-confidence label assignments were annotated, resulting in relatively sparse labels and leaving many unlabeled voxels near the thalamic boundaries (Fig. 1(b)).

2.2   Data Processing and Computation

We preprocess MPRAGE and FGATIR following the procedure in [9] to generate PD and T_1 maps, and multi-TI images. MPRAGE and FGATIR are co-registered to MNI space using ANTS [2]. Both images then undergo N4 Bias Field Correction and White Matter Mean Normalization. Notably, these images are processed jointly using a harmonic bias field and a common scaling factor determined from a white matter mask derived from MPRAGE, to ensure accurate maps estimation.

In a T1-weighted MRI, a voxel v's intensity $I(v)$ can be modeled as:

$$I(v) = \mathrm{PD}(v)[1 - 2e^{(-\mathrm{TI}/T_1(v))} + e^{(-\mathrm{TR}/T_1(v))}], \tag{1}$$

where $PD(v)$ is the voxel's proton density, $T_1(v)$ the T1 relaxation time, TI the inversion time, and TR the repetition time. As described in [9], by acquiring MPRAGE and FGATIR with the same parameters but different TIs, we can determine each voxel's PD and T_1. The PD and T_1 maps enables further computation of T1-weighted images at any chosen TI. In Fig. 2, example multi-TI images show how TI variations affect tissue contrast and highlight different features.

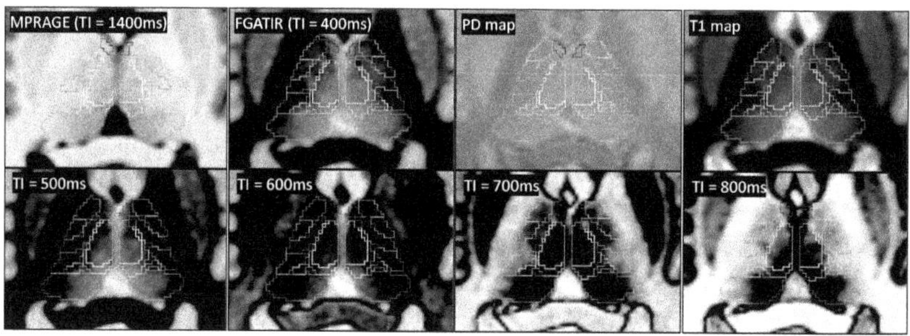

Fig. 2. MPRAGE, FGATIR, PD and T_1 maps, and example multi-TI images.

2.3 Identify the Optimal Inputs

We aim to identify the most effective input modalities for thalamic nuclei segmentation through a two-stage process. Stage 1 selects the most informative multi-TI images. Stage 2 compares structural sequences (MPRAGE, FGATIR) against quantitative maps (PD, T_1 maps) and the selected multi-TI images.

Stage 1: Selecting informative multi-TI images

(1) Segmentation Model and Input Images: A 3D U-Net [4] with Dice loss is adopted to segment the thalamus into 14 classes, including 13 nuclei (or nuclear groups) and background. We trained the model on 53 3D images per subject. These include: 1) 51 multi-TI images, computed at TIs from 400ms to 1400ms in 20ms increments. This range captures meaningful contrast differences in thalamic structures, with 20ms as the smallest visually discernible step. 2) MPRAGE and FGATIR, corresponding to the 1400 ms and 400 ms TIs, respectively. While the multi-TI images are derived from quantitative maps, MPRAGE and FGATIR are directly acquired. All 53 images are treated as distinct input contrasts.

(2) Gradient Analysis with Monte Carlo Dropout: To quantify the contribution of each input image, we employ gradient-based saliency analysis during inference. Specifically, we compute the gradients of the model's output probability for each class with respect to input images. Unlike traditional pixel-wise saliency maps that focus on individual voxel contributions, our approach captures channel-wise importance. Larger gradients indicate stronger influence on

the model's prediction, while smaller gradients reflect weaker contributions. This method is interpretable and computationally efficient, as it leverages the network's existing backpropagation mechanism to reveal each input's relevance without adding extra computational overhead. However, single gradient computations can be sensitive to factors such as model architecture, random initialization, and inference-time stochasticity. These sources of variability may lead to unstable estimates of input importance. To address this, we apply Monte Carlo (MC) Dropout during inference, performing multiple forward-backward passes with dropout enabled. This simulates an ensemble of different "sub-models" without training separate networks, efficiently capturing variability across runs. Gradients computed across multiple backward passes are then integrated to obtain robust estimates of each image's contribution, guiding the selection of the most informative inputs.

(3) Overall Importance Score (OIS): We define the OIS to summarize the gradient information and rank the image contributions.

$$\text{OIS}_i = \frac{1}{SM} \sum_{(c,s,m)\in\mathcal{X}} \left| \nabla_{I_{i,s}} \sum_{v\in\mathcal{V}_s} f_\theta^{(m)}(I_{.,s})[c,v] \right|, \quad \text{where } \mathcal{X} = C \times S \times M$$

where:

- S, C, M: Number of subjects, segmentation classes, and test-time MC dropout runs per subject, respectively.
- $\mathcal{V}_s$: The set of 3D voxel coordinates for subject s. All co-registered input images share the same spatial grid ($|\mathcal{V}_s| = H \times W \times L$).
- $I_{i,s}$: The i-th input image for subject s.
- $I_{.,s}$: All input images for subject s, stacked as a multi-channel input.
- $f_\theta^{(m)}$: Segmentation model with dropout applied during the m-th MC run.
- $f_\theta^{(m)}(I_{.,s})[c,v]$: The predicted probability that voxel v belongs to class c.
- $\nabla_{I_{i,s}}$: Gradient operator with respect to input image $I_{i,s}$.

The OIS quantifies model dependence by measuring how sensitively the output responds to changes in each input image. We compute the gradient by first summing predicted probabilities over all voxels for a given class, then backpropagating this scalar with respect to each input. While gradients are computed voxel-wise, we aggregate them across voxels to obtain a single scalar importance score per input image. We use absolute gradient values, focusing on how strongly the image affects the output rather than direction of change. Aggregating across segmentation classes, subjects, and MC dropout runs, OIS identifies inputs that are broadly informative, robust to inter-subject variability, and stable under different model configurations. Higher OIS values indicate stronger influence on segmentation, while lower values suggest minimal relevance.

Stage 2: Input Modality Comparison
Building on Stage 1 results, we now compare the selected multi-TI subsets against structural sequences and quantitative maps to determine the most effective input modality. The candidate multi-TI subsets identified by OIS ranking

are combined with MPRAGE, FGATIR, T_1 maps, and PD maps to form multiple candidate input configurations. For each configuration, we train a separate 3D U-Net [4] on volumetric data from scratch using Dice loss and the same training protocol to ensure fair comparison. Configuration details are provided in Sect. 3.3.

3   Experiments and Results

3.1   Experimental Setup

All models in both Stage 1 and Stage 2 share the same training protocol. The segmentation model was trained with LeakyReLU activation, instance normalization, and Adam optimization (lr = 0.001, weight decay=0.0001). The learning rate decreases by 10% after 5 stagnant epochs, and early stopping occurs after 15 epochs without validation improvement. Data augmentation including left-right flipping, affine transformations (scaling, rotation, translation), and cropping to 96×96×96 are applied to boost robustness. We perform 8-fold cross-validation on 24 subjects, splitting data into training, validation, and testing with a 19:2:3 ratio per fold, ensuring each subject is tested once. Since the ground truth labels are sparse, we use true positive rate (TPR) rather than Dice score as our evaluation metric to avoid unfairly penalizing correctly segmented but unlabeled voxels.

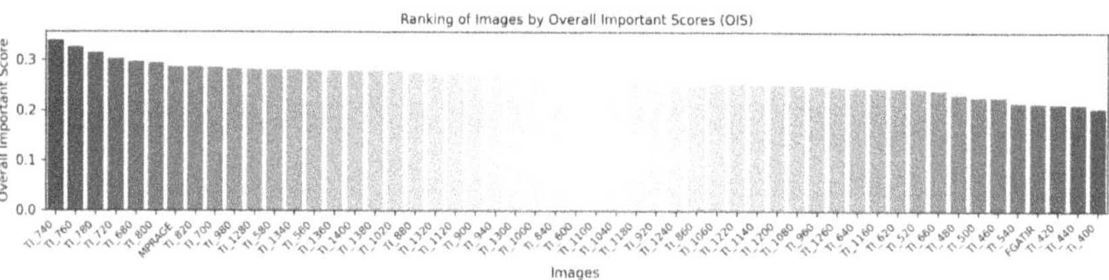

Fig. 3. Ranking of images by Overall Importance Score (OIS). The x-axis represents input images, while the y-axis shows their corresponding OIS values.

3.2   Selecting the Most Informative Multi-TI Images

As detailed in Sect. 2.3, to identify the most informative multi-TI images, we performed gradient-based saliency analysis with test-time Monte Carlo (MC) Dropout using a 0.1 dropout rate. In 8-fold cross-validation, 24 subjects each served as the test set once and underwent 100 forward-backward passes. We then computed the Overall Importance Score (OIS) to quantify each input image's contribution. Figure 3 ranks the images by their OIS, where high scores indicate strong and stable contributions. Based on this ranking, we selected three candidate configurations for further evaluation: 1) the top TI at 740 ms, 2) the top two TIs at 740 and 760 ms, and 3) the top four TIs at 740, 760, 780, and 720 ms.

3.3   Comparative Study of Input Configurations

Building on the selected multi-TI subsets from Sect. 3.2, we now compare them against structural sequences and quantitative maps across nine configurations: Config$_1$: MPRAGE, Config$_2$: FGATIR, Config$_3$ MPRAGE+FGATIR, Config$_4$: PD+T_1 maps, Config$_5$: T_1 map, Config$_6$: 51 multi-TI images (400–1400 ms, in 20 ms steps), Config$_7$: the top multi-TI image at 720ms, Config$_8$: the top two at 740 and 760 ms, Config$_9$: the top four at 720, 740, 760, and 780 ms.

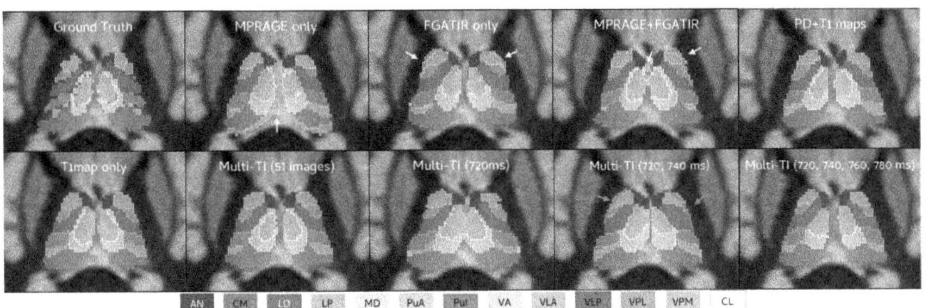

Fig. 4. Qualitative comparison of different input configurations.

We evaluate these configurations both quantitatively and qualitatively. Quantitatively, we computed the TPR for each thalamic nucleus and the volume-weighted average (VWA) across 13 nuclei. The mean and standard deviation (SD) of both per-nucleus TPR and VWA across 24 subjects are shown in Table 1. Config$_8$ (top2 multi-TIs) achieves the highest VWA of 80.58±3.80%, indicating that carefully selected multi-TI subsets capture essential features more effectively than MPRAGE and FGATIR, or even the full multi-TI set. Config$_5$ (T_1map) (highlighted in blue) performs comparably well, suggesting that T_1 maps alone are sufficient for accurate segmentation. Qualitatively, as shown in Fig. 4, MPRAGE and FGATIR-based setups often misplace the small PuA nucleus (1.3% of thalamus), marked by the yellow arrows. While Config$_8$ (top2 multi-TIs) yields the best TPR overall, it slightly over-segments the thalamic border, marked by the red arrows. Config$_4$ (PD+T_1 maps) and Config$_5$ (T_1map) offer high-quality segmentations with minimal difference, implying PD maps contribute little beyond T_1 maps.

For statistical validation, we use a Wilcoxon signed-rank test to compare each configuration with Config$_5$ (T_1map) on the per-nucleus TPR and VWA. A Holm–Bonferroni correction was used to control for multiple comparisons, with significance set at $p < 0.05$. In Table 1, ↑↑ and ↓↓ indicate significantly higher or lower TPR than Config$_5$ (T_1map). Notably, there is no significant difference between Config$_8$ (top2 multi-TIs) and Config$_5$ (T_1map), reinforcing that T_1 map alone suffices. Config$_5$ (T_1map) performs similarly to Config$_4$ (PD+T_1 maps), even surpassing it in VPM, confirming that PD maps provide minimal additional benefit. Moreover, Config$_5$ (T_1map) outperforms most other experiments across multiple nuclei and VWA, establishing it as the most effective and efficient input.

Table 1. Quantitative comparison of input configurations. Mean $\pm$ SD of TPR per nucleus and the volume-weighted average (VWA) across 24 subjects, reported in percentages. Fractions under each nucleus indicates its proportional size in the thalamus. $\uparrow\uparrow$ denotes values significantly higher than Config_5 (T_1map), $\downarrow\downarrow$ indicates significantly lower, and no arrow indicates no significant difference.

	AN 3.9%	CM 3.1%	LD 1.5%	LP 5.5%	MD 13.3%	PuA 1.3%	Pul 27.6%	VA 8.1%	VLA 3.4%	VLP 16.2%	VPL 6.3%	VPM 1.8%	CL 7.4%	VWA
MPR	72.47 ± 9.71	79.70 ± 5.35	64.50 ± 17.38	76.72 ± 6.95	85.52 ± 3.91	0.00 ±↓↓ 0.00	85.35 ± 4.47	79.80 ± 6.26	75.57 ± 9.76	83.72 ± 4.08	80.40 ± 7.49	67.23 ± 8.46	**80.40** ±↑↑ **5.84**	79.07 ± 2.64
FGT	73.21 ± 11.62	77.67 ±↓↓ 6.22	63.06 ± 16.27	80.43 ± 8.03	82.79 ±↓↓ 5.11	42.46 ± 34.47	**86.67** ± **4.58**	78.32 ± 9.47	76.33 ± 9.67	82.99 ± 6.71	**80.61** ± **6.87**	69.54 ± 9.73	61.26 ± 8.03	79.71 ± 2.94
M+F	75.91 ± 9.62	78.63 ± 7.78	63.26 ± 16.46	77.64 ± 7.39	87.02 ± 6.42	8.69 ±↓↓ 23.03	85.15 ± 5.50	81.27 ± 6.33	75.23 ± 7.05	82.57 ± 4.39	78.86 ± 6.58	23.65 ±↓↓ 34.90	62.79 ± 8.62	78.62 ±↓↓ 3.36
PD+T_1	71.57 ± 11.80	77.15 ± 7.61	65.06 ± 17.01	**81.06** ± **6.06**	87.24 ± 6.11	60.23 ± 25.75	86.06 ± 5.23	82.45 ± 5.32	**76.84** ± **11.85**	82.84 ± 4.03	79.37 ± 5.53	58.61 ±↓↓ 24.95	61.12 ± 7.29	80.34 ± 2.89
T_1map	76.44 ± 10.09	**81.68** ± **6.91**	63.63 ± 17.65	78.90 ± 5.94	87.76 ± 5.82	40.65 ± 32.84	85.35 ± 4.61	76.27 ± 13.15	76.69 ± 7.22	82.79 ± 4.54	80.08 ± 5.72	**71.15** ± **7.68**	64.66 ± 6.30	80.20 ± 2.86
TI(51)	73.79 ± 10.48	73.18 ± 13.32	48.31 ± 23.51	76.73 ± 8.08	85.06 ± 5.53	23.32 ± 31.76	86.25 ± 4.44	80.76 ± 5.16	75.79 ± 8.72	82.89 ± 3.25	80.33 ± 5.52	68.58 ± 8.85	67.29 ± 7.19	79.66 ± 2.70
TI(top1)	53.12 ±↓↓ 18.09	73.96 ±↓↓ 14.41	**66.33** ± **20.37**	78.67 ± 8.00	54.81 ±↓↓ 17.21	**68.70** ±↑↑ **9.28**	83.78 ± 6.52	52.84 ±↓↓ 17.26	74.21 ± 10.75	77.98 ± 14.06	73.91 ± 14.21	68.63 ± 8.74	59.04 ± 14.03	70.90 ±↓↓ 4.22
TI(top2)	**77.09** ± **10.12**	79.73 ± 8.25	58.08 ± 17.60	77.75 ± 8.20	**89.84** ± **4.21**	66.83 ± 10.66	84.04 ± 6.13	**82.98** ± **4.78**	72.23 ± 11.18	81.73 ± 4.61	78.18 ± 5.99	67.92 ± 8.83	67.83 ± 8.17	**80.58** ± **3.80**
TI(top4)	74.06 ± 13.16	54.83 ±↓↓ 19.48	65.79 ± 22.49	78.52 ± 9.99	84.84 ± 6.35	49.81 ± 25.98	85.22 ± 5.29	81.82 ± 5.69	72.44 ± 8.78	**84.13** ± **4.87**	77.61 ± 6.84	68.58 ± 11.14	69.05 ± 6.40	78.87 ± 3.23

4   Discussions and Conclusions

Our experiments show that the essential tissue contrast captured by multi-TI images is inherently encoded in the T_1 map. Since multi-TI images are derived from the T_1 map, direct segmentation using the T_1 map reduces preprocessing demands while preserving key information. Additionally, multi-TI images must be carefully selected, as blindly incorporating all 51 multi-TI images does not outperform the T_1map-only approach, suggesting that an unfiltered multi-TI strategy is less effective than simply leveraging the T_1 map itself. Moreover, PD maps offer no added value beyond T_1 maps, reinforcing T_1 maps as the primary source of contrast. Although some T_1-map features are not visually apparent, computational models can extract meaningful structures beyond human perception. Given this, is there still value in computing multi-TI images? We argue yes, as they offer enhanced visual contrast that supports manual delineation.

A practical concern is the availability of high-quality T_1 maps. When not directly accessible, they can be estimated from MPRAGE or FGATIR, acquired via MP2RAGE [17], or synthesized from MPRAGE using learning-based methods [12]. Our approach has several limitations: First, the dataset size is relatively small, though we mitigate this with 8-fold cross-validation and statistical evaluation. Second, external validation remains challenging because most public datasets (e.g., ADNI, HCP, BLSA, IXI) lack key contrasts such as FGATIR or T_1 maps, limiting their direct use for evaluation. Future work will focus on synthesizing T_1 maps in these datasets and training more generalizable models. For datasets without manual labels, we plan to explore semi-supervised or pseudo-labeling strategies. Third, we do not explore all imaging modalities; our comparisons focus mainly on T1w-based inputs. However, modalities such as T2w, diffusion MRI, and structural connectivity are excluded intentionally. T2w images are rarely used for thalamic segmentation due to limited intra-thalamic contrast. Diffusion and connectivity features capture microstructural information that may not structurally align with T1w anatomy. Since our manual labels are defined on T1w images, evaluating non-T1w modalities against them would inherently disadvantage these inputs and result in unfair comparisons.

In conclusion, in this paper, we systematically evaluated structural sequences (MPRAGE, FGATIR), quantitative maps (PD, T_1 maps), and multi-TI images for thalamic nuclei segmentation. To select the most informative multi-TI images, we introduced a novel approach that integrates gradient-based saliency analysis with test-time Monte Carlo dropout and proposed the Overall Importance Score to quantify each image's contribution. This evaluation framework is generalizable to any input selection or feature attribution task. Moreover, our findings show that T_1map-based segmentation offers the best balance of accuracy and efficiency, making it the most practical choice for thalamic nuclei segmentation.

Acknowledgment and Disclaimer. This research was supported in part by the Intramural Research Program of the National Institutes of Health (NIH). The contributions of the NIH authors were made as part of their official duties as NIH federal employees, are in compliance with agency policy requirements, and are considered Works of the United States Government. However, the findings presented in this paper are those of the authors and do not necessarily reflect the views of the NIH or the U.S. Department of Health and Human Services. The work is also supported in part by NIH through the National Institute of Neurological Disorders and Stroke grant R01-NS105503 (PIs: Zhuo & Prince).

References

1. Aggleton, J.P., Pralus, A., Nelson, A.J., Hornberger, M.: Thalamic pathology and memory loss in early Alzheimer's disease: moving the focus from the medial temporal lobe to papez circuit. Brain **139**(7), 1877–1890 (2016)
2. Avants, B.B., Tustison, N., Song, G.: Advanced normalization tools (ANTS). Insight J. **2**(365), 1–35 (2009)
3. Braak, H., Braak, E.: Alzheimer's disease affects limbic nuclei of the thalamus. Acta Neuropathol. **81**(3), 261–268 (1991)

4. Çiçek, Ö., Abdulkadir, A., Lienkamp, S.S., Brox, T., Ronneberger, O.: 3D U-Net: learning dense volumetric segmentation from sparse annotation. In: Ourselin, S., Joskowicz, L., Sabuncu, M.R., Unal, G., Wells, W. (eds.) MICCAI 2016. LNCS, vol. 9901, pp. 424–432. Springer, Cham (2016). https://doi.org/10.1007/978-3-319-46723-8_49

5. Deoni, S.C., Josseau, M.J., Rutt, B.K., Peters, T.M.: Visualization of thalamic nuclei on high resolution, multi-averaged T1 and T2 maps acquired at 1.5 T. Human Brain Mapping **25**(3), 353–359 (2005)

6. Deoni, S.C., Rutt, B.K., Parrent, A.G., Peters, T.M.: Segmentation of thalamic nuclei using a modified k-means clustering algorithm and high-resolution quantitative magnetic resonance imaging at 1.5 T. NeuroImage **34**(1), 117–126 (2007)

7. Duan, Y., Xi, Y.: Thalamus segmentation from diffusion tensor magnetic resonance imaging. Int. J. Biomed. Imaging **2007**(1), 090216 (2007)

8. Eaton-Rosen, Z., Bragman, F., Bisdas, S., Ourselin, S., Cardoso, M.J.: Towards safe deep learning: accurately quantifying biomarker uncertainty in neural network predictions. In: Frangi, A.F., Schnabel, J.A., Davatzikos, C., Alberola-López, C., Fichtinger, G. (eds.) MICCAI 2018. LNCS, vol. 11070, pp. 691–699. Springer, Cham (2018). https://doi.org/10.1007/978-3-030-00928-1_78

9. Feng, A., Bian, Z., Dewey, B.E., Colinco, A.G., Zhuo, J., Prince, J.L.: RATNUS: Rapid, Automatic Thalamic Nuclei Segmentation using Multimodal MRI inputs. ArXiv Preprint ArXiv:2409.06897 (2024)

10. Feng, A., et al.: Label propagation via random walk for training robust thalamus nuclei parcellation model from noisy annotations. In: 2023 IEEE 20th International Symposium on Biomedical Imaging (ISBI), pp. 1–5. IEEE (2023)

11. Gal, Y., Ghahramani, Z.: Dropout as a bayesian approximation: Representing model uncertainty in deep learning. In: ICML, pp. 1050–1059. PMLR (2016)

12. Hays, S.P., et al.: Revisiting registration-based image synthesis: a focus on unsupervised mr image synthesis. In: Medical Imaging 2024: Image Processing, vol. 12926, pp. 257–265. SPIE (2024)

13. Iglesias, J.E., Insausti, R., Lerma-Usabiaga, G., Bocchetta, M., Van Leemput, K., Greve, D.N., Van der Kouwe, A., Fischl, B., Caballero-Gaudes, C., Paz-Alonso, P.M., et al.: A probabilistic atlas of the human thalamic nuclei combining ex vivo MRI and histology. Neuroimage **183**, 314–326 (2018)

14. Johansen-Berg, H., Behrens, T.E., Sillery, E., Smith, S.M., Matthews, P.M.: Functional-anatomical validation and individual variation of diffusion tractography-based segmentation of the human thalamus. Cereb. Cortex **15**(1), 31–39 (2005)

15. Kawahara, J., et al.: BrainNetCNN: convolutional neural networks for brain networks; towards predicting neuro development. Neuroimage **146**, 1038–1049 (2017)

16. Kumar, R., et al.: Long-term follow-up of thalamic deep brain stimulation for essential and parkinsonian tremor. Neurology **61**(11), 1601–1604 (2003)

17. Marques, J.P., Kober, T., Krueger, W., Gruetter, R.: MP2RAGE, a self bias-field corrected sequence for improved segmentation and T1-mapping at high field. Neuroimage **49**(2), 1271–1281 (2010)

18. Morel, A., Magnin, M., Jeanmonod, D.: Multiarchitectonic and stereotactic atlas of the human thalamus. Journal of Comparative Neurology **387**(4), 588–630 (1997)

19. Mugler, J.P., III., Brookeman, J.R.: Three-dimensional magnetization-prepared rapid gradient-echo imaging (3D MPRAGE). Magn. Reson. Med. **15**(1), 152–157 (1990)

20. Nair, T., Precup, D., Arnold, D.L., Arbel, T.: Exploring uncertainty measures in deep networks for multiple sclerosis lesion detection and segmentation. Med. Image Anal. **59**, 101557 (2020)
21. Najdenovska, E., et al.: In-vivo probabilistic atlas of human thalamic nuclei based on diffusion-weighted magnetic resonance imaging. Sci. Data **5**(1), 1–11 (2018)
22. O'Muircheartaigh, J., et al.: Clustering probabilistic tractograms using independent component analysis applied to the thalamus. Neuroimage **54**(3), 2020–2032 (2011)
23. Papavassiliou, E., et al.: Thalamic deep brain stimulation for essential tremor: relation of lead location to outcome. Neurosurgery **62**, SHC–884 (2008)
24. Rispal-Padel, L., Massion, J., Grangetto, A.: Relations between the ventrolateral thalamic nucleus and motor cortex and their possible role in the central organization of motor control. Brain Res. **60**(1), 1–20 (1973)
25. Saranathan, M., Iglehart, C., Monti, M., et al.: In vivo high-resolution structural MRI-based atlas of human thalamic nuclei. Sci. Data **8**(1), 275 (2021)
26. Selvaraju, R.R., Cogswell, M., Das, A., Vedantam, R., et al.: Grad-CAM: visual Explanations from Deep Networks via Gradient-based Localization. In: Proceedings of the IEEE International Conference on Computer Vision, pp. 618–626 (2017)
27. Sherman, S.M., Guillery, R.W.: Exploring the thalamus and its role in cortical function. MIT Press (2006)
28. Sherman, S.M., Guillery, R.W.: Exploring the Thalamus. Elsevier (2001)
29. Sommer, M.A.: The role of the thalamus in motor control. Curr. Opin. Neurobiol. **13**(6), 663–670 (2003)
30. Su, J.H., Thomas, F.T., Kasoff, W.S., Tourdias, T., Choi, E.Y., et al.: Thalamus Optimized Multi Atlas Segmentation (THOMAS): fast, fully automated segmentation of thalamic nuclei from structural MRI. Neuroimage **194**, 272–282 (2019)
31. Sudhyadhom, A., Haq, I.U., et al.: A high resolution and high contrast MRI for differentiation of subcortical structures for DBS targeting: the Fast Gray Matter Acquisition T1 Inversion Recovery (FGATIR). Neuroimage **47**, T44–T52 (2009)
32. Sundararajan, M., Taly, A., Yan, Q.: Axiomatic attribution for deep networks. In: International Conference on Machine Learning, pp. 3319–3328. PMLR (2017)
33. Tourdias, T., Saranathan, M., et al.: Visualization of intra-thalamic nuclei with optimized white-matter-nulled MPRAGE at 7 T. Neuroimage **84**, 534–545 (2014)
34. Traynor, C.R., Barker, G.J., Williams, S.C., Richardson, M.P.: Segmentation of the thalamus in MRI based on T1 and T2. Neuroimage **56**(3), 939–950 (2011)
35. Umapathy, L., Keerthivasan, M.B., Bilgin, A., Saranathan, M.: Convolutional neural network based frameworks for fast automatic segmentation of thalamic nuclei from native and synthesized contrast structural MRI. Neuroinformatics, pp. 1–14 (2022)
36. Wiegell, M.R., Tuch, D.S., et al.: Automatic segmentation of thalamic nuclei from diffusion tensor magnetic resonance imaging. Neuroimage **19**(2), 391–401 (2003)
37. Yan, C., et al.: Segmenting thalamic nuclei from manifold projections of multi-contrast MRI. In: Medical Imaging 2023: Image Processing, vol. 12464, pp. 727–734. SPIE (2023)

Radiomic Fingerprints for Knee MR Image Assessment

Yaxi Chen[1,2(✉)] , Simin Ni[3] , Shaheer U. Saeed[2,4] , Aleksandra Ivanova[3] ,
Rikin Hargunani[5] , Jie Huang[1] , Chaozong Liu[3,5] , and Yipeng Hu[2,4]

[1] Mechanical Engineering Department, University College London, London, UK

[2] UCL Hawkes Institute, University College London, London, UK
yaxi.chen.20@uci.ac.uk

[3] Institute of Orthopaedic and Musculoskeletal Science, University College London,
Royal National Orthopaedic Hospital, Stanmore, UK

[4] Department of Medical Physics and Biomedical Engineering,
University College London, London, UK

[5] Royal National Orthopaedic Hospital, Stanmore, UK

Abstract. Accurate interpretation of knee MRI scans relies on expert clinical judgment, often with high variability and limited scalability. Existing radiomic approaches use a fixed set of radiomic features (the "signature"), selected at the population level and applied uniformly to all patients. While interpretable, these signatures are often too constrained to represent individual pathological variations. As a result, conventional radiomic-based approaches are found to be limited in performance, compared with recent end-to-end deep learning (DL) alternatives without using interpretable radiomic features.

We argue that the individual-agnostic nature in current radiomic selection is not central to its interpretability, but is largely responsible for the poor generalization in our application. Here, we propose a novel radiomic fingerprint framework, in which a radiomic feature set (the "fingerprint") is dynamically constructed for each patient, selected by a DL model. Unlike the existing radiomic signatures, our fingerprints are derived on a per-patient basis by predicting the feature relevance in a large radiomic feature pool, and selecting only those that are predictive of clinical conditions for individual patients. The radiomic-selecting model is trained simultaneously with a low-dimensional (considered relatively explainable) logistic regression for downstream classification.

We validate our methods across multiple diagnostic tasks including general knee abnormalities, anterior cruciate ligament (ACL) tears, and meniscus tears, demonstrating comparable or superior diagnostic accuracy relative to state-of-the-art end-to-end DL models. Perhaps more importantly, we show that the interpretability inherent in our approach facilitates meaningful clinical insights and potential biomarker discovery, with detailed discussion, quantitative and qualitative analysis of real-world clinical cases to evidence these advantages. The codes are available at: https://github.com/YaxiiC/RadiomicsFingerprint.git

© The Author(s), under exclusive license to Springer Nature Switzerland AG 2026
M. Reyes et al. (Eds.): iMIMIC 2025, LNCS 16464, pp. 64–74, 2026.
https://doi.org/10.1007/978-3-032-17611-0_7

Keywords: Knee Joint · Radiomics · MRI · Clinical Interpretability

1   Introduction

Common and clinically significant knee injuries, such as anterior cruciate ligament (ACL) tears and meniscal tears, are routinely evaluated using MRI [11] as gold standard due to its exceptional capability to visualize detailed anatomical structures and pathological changes non-invasively. MRI provides critical insights into the precise location, depth, and patterns of tissue injuries, making it indispensable in clinical decision-making and treatment planning. Beyond detection, MRI enables detailed assessment of injury characteristics, including the distinction between partial and complete ACL tears, as well as the classification of meniscal damage, which guides decisions between repair and resection [1]. These insights are critical for selecting between conservative management, such as physiotherapy, and surgical options, including ACL reconstruction or meniscal repair and meniscectomy. With increasing imaging demand, automated analysis tools are increasingly needed to support timely and informed clinical decision making.

Existing approaches for automated knee MRI assessment generally fall into two broad paradigms: classical radiomics and deep learning (DL). Radiomics involves the extraction of hand-crafted quantitative features from regions of interest (ROI) in the images, capturing characteristics such as intensity distributions, texture patterns, shape descriptors, and wavelet coefficients [14]. Such radiomic models are valued for their interpretability – each feature has a well-defined meaning (intensity, roughness, size, etc.) that can be understood by clinicians. A radiologist can potentially relate a feature's value to an aspect of the anatomy or pathology, making the decision process more transparent. The trade-off, however, is that radiomics use a fixed, hand-crafted feature set that may not capture the full complexity of imaging patterns, and these simpler models often yield lower diagnostic accuracy than state-of-the-art DL in practice [2,8]. Additionally, radiomic features can be sensitive to imaging protocol variations, raising generalizability concerns if not carefully standardized.

In contrast, DL methods learn their own task-specific features directly from images, typically achieving higher accuracy in various downstream tasks. Most previous studies in this area have utilized end-to-end convolutional neural networks (CNNs). For instance, Bien et al. [3] utilized an end-to-end classifier and Tsai et al. [12] applied an a lightweight efficiently layered network (ELNet). These DL models demonstrated high diagnostic performance for knee pathology detection, often approaching expert-level sensitivity and specificity. The decision mechanism of a CNN is distributed across many abstract features, making it difficult to interpret which image characteristics led to a given prediction. While methods like class activation mapping (CAM) [3] have been used to enhance interpretability, a major limitation of these DL models remains their "black box" nature. The models generally lack the transparent feature-by-feature reasoning that radiomics offers. The lack of transparency and thus human understanding of how image-level features are learned and decisions are made remains an interesting challenge. Numerous studies in explainable AI emphasize this concern and call for transparent models in high-stakes domains [4].

Recent studies have suggested that integrating hand-crafted features with deep features or using hybrid models can yield more robust results in medical imaging tasks [7]. This gap - between transparent but rigid radiomics and accurate but inscrutable DL - has motivated our proposed patient-specific radiomic fingerprint framework. Instead of relying on a static, one-size-fits-all feature set, our method dynamically tailors the radiomic features to each individual case via a neural network-based feature selection mechanism. Specifically, we employ a feature-weighting neural network that analyzes the patient's MRI and assigns an importance weight (or selection probability) to each feature in a large pool of candidate radiomic features. In this way, the model learns to select an optimal subset of features for each patient – effectively constructing a personalized radiomic fingerprint on the fly, tuned to the particular imaging characteristics of that patient's knee. These selected features then serve as inputs to a downstream logistic regression model, which produces the final classification (e.g., presence or absence of an ACL tear). The feature-weighting network and the logistic regression classifier are trained jointly in an end-to-end fashion, under supervision of the known condition labels. This integration ensures that the network learns to choose feature subsets that maximize the downstream diagnostic performance, while the logistic regression provides a simple, interpretable mapping from those features to the prediction. This study builds on the previous work [5], which integrated radiomics with reconstructed patient-specific healthy baselines and emphasized visual interpretability. In contrast, this work focuses on maximising diagnostic performance while exploring interpretability based solely on radiomic fingerprints. We expand the feature pool to include both first- and higher-order texture features to build a comprehensive radiomic descriptors, enhancing fingerprint expressiveness and diagnostic utility.

Our approach achieves the dual goals of preserving interpretability and improving clinical performance in automated knee MRI diagnosis. The DL–guided dynamic feature selection empowers the model with patient-specific flexibility and enhanced accuracy. The result is a tailored radiomic fingerprint for each patient that is both explainable and predictive. We summarize our key contributions as follows: 1) A patient-specific radiomic fingerprint approach that improves interpretability and performance; 2) Validation demonstrating comparable or superior diagnostic accuracy compared to DL methods in multiple knee MRI classification tasks; and 3) An open-source implementation enabling reproducibility and future advancements in interpretable medical imaging.

2   Methods

Our method is built around the concept of radiomic fingerprint - a subject-specific, dynamically constructed representation derived from a large pool of radiomic features. The radiomic fingerprint in our framework is computed by predicting a dense relevance vector over a comprehensive set of hand-crafted radiomic features extracted from multi-scale, multi-view subregions of the image. These feature weight are modeled as continuous-valued probabilities, which are thresholded at inference time to yield a binary selection for individual radiomic features. The final fingerprint is therefore a sparse, interpretable, yet highly discriminative set of radiomic features tailored to each patient. Because selection occurs over a rich feature pool, the effective feature space of the fingerprint is significantly larger than that of any fixed radiomic signature, supporting greater expressivity and finer-grained diagnostic encoding.

The following subsections describe the feature extraction process, the neural network architecture used for relevance prediction, the construction of the fingerprint vector, and the downstream classification procedure.

Feature Extraction: Given an image ROI $\mathbf{x} \in \mathbb{R}^{H \times W \times D}$, where H, W, and D denote its spatial dimensions. To represent the spatially local variation, the ROI is further divided into J non-overlapping patches. From each patch, a collection of 3D radiomic features are computed independently on three different image volumes: axial, sagittal, and coronal views. Further details on the definitions of radiomic features, ROIs, and registration aligning them from the three sequences are described in Sect. 3.1.

Let $f_{j,k}^{(v)}$ denote the kth feature extracted from the jth patch in v^{th} image sequence, where $k = 1, \dots, K$, $j = 1, \dots, J$ and $v \in \{1,2,3\}$. Thus, the feature vector $\mathbf{f}_j$ for the jth patch:

$$\mathbf{f}_j = [f_{j,1}^{(1)}, ..., f_{j,K}^{(1)}, f_{j,1}^{(2)}, ..., f_{j,K}^{(2)}, f_{j,1}^{(3)}, ..., f_{j,K}^{(3)}]^{\top} \in \mathbb{R}^{3K}$$

Now, the full feature vector $\mathbf{f}$ assembles $\mathbf{f}_j$ vectors from all J patches:

$$\mathbf{f} = [\mathbf{f}_1^{\top}, \mathbf{f}_2^{\top}, ..., \mathbf{f}_J^{\top}]^{\top} \in \mathbb{R}^{3JK}$$

For notational brevity, the full feature vector $\mathbf{f}$ is henceforth denoted with features f_i using a single subscript $i = 1, ..., 3JK$:

$$\mathbf{f} = [f_1, f_2, ..., f_{3JK}]^{\top}$$

Dynamic Feature Selection via Neural Network: A neural network $G_\omega(\mathbf{x})$, parameterized by ω, takes the entire ROI x as input and predicts a relevance score vector $\mathbf{q}$ for all computed radiomic features $\mathbf{f}$:

$$\mathbf{q} = G_\omega(\mathbf{x}) = [q_1, q_2, ..., q_{3JK}]^{\top} \in \mathbb{R}^{3JK}$$

Each scalar weight $q_i \in [0, 1]$ quantifies the importance of individual features f_i, with respect to the downstream task. During training, the relevance scores $\mathbf{q}$ are applied to the corresponding features via element-wise multiplication:

$$\mathbf{f}^{\mathrm{w}} = \mathbf{f} \odot \mathbf{q} = [f_1 \cdot q_1, f_2 \cdot q_2, \dots, f_{3JK} \cdot q_{3JK}]^{T}.$$

Downstream Classification: The weighted feature vector $\mathbf{f}^{w}$ represents the soft assignment of individual features. Normalisation is not applied to allow flexibility in determining empirical threshold values (Sect. 3.1), when taking into account inference-time application-specific requirement, e.g. size of selected feature set. This vector is passed directly to a clinical task classifier R_ψ, parameterized by ψ, to predict the final diagnostic class probabilities $\hat{\mathbf{c}} = R_\psi(\mathbf{f}^{w})$.

This classifier is designed as a low-dimensional function without higher-degree terms, for its interpretability, such as the logistic regression used in this study, and is commonly adopted in classical radiomic signature algorithms for the same considerations.

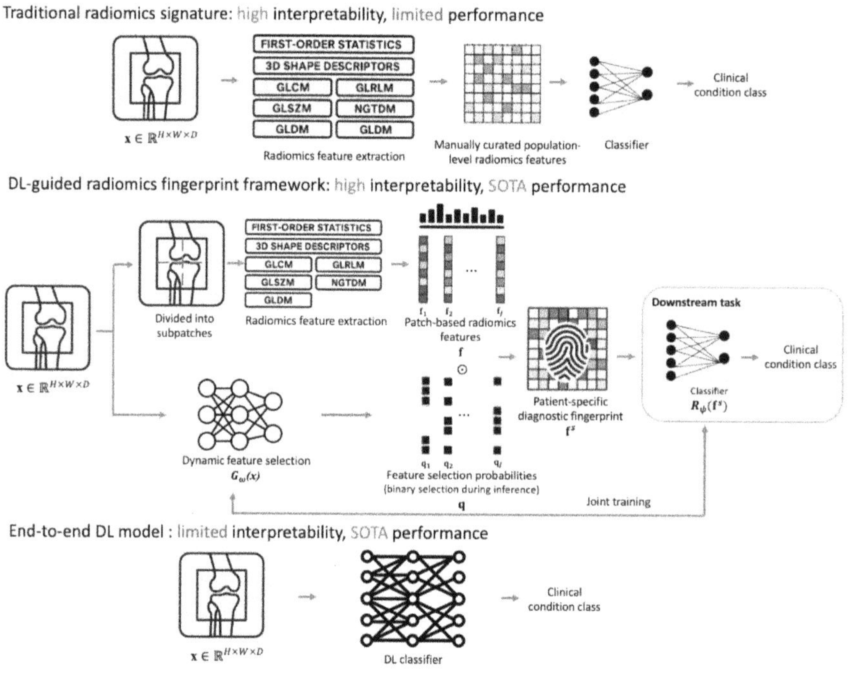

Fig. 1. Comparison between the traditional radiomic signature approaches, our proposed radiomic fingerprint framework, and example end-to-end DL models.

Training: To summarise, features $\mathbf{f}$ are extracted from all patches across all three view-differing MR sequences. The feature-weighting network G_ω takes the full ROI as input and predicts the corresponding weight $\mathbf{q}$. The weighted feature vector $\mathbf{f}^{w} = \mathbf{f} \odot \mathbf{q}$ is the input of the classifier R_ψ. The parameters of both the network and the classifier are optimized jointly by minimizing a cross-entropy loss $\mathcal{L}(\hat{\mathbf{c}}, \tilde{\mathbf{c}})$ over the training set, with available class labels $\tilde{\mathbf{c}}$:

$$(\omega^{*}, \psi^{*}) = \arg\min_{\omega, \psi} \sum \mathcal{L}(R_\psi(\mathbf{f} \odot G_\omega(\mathbf{x})), \tilde{\mathbf{c}}).$$

Feature Selection at Inference: At inference time, the continuous feature weights predicted by the feature-weighting network G_ω are thresholded to obtain a binary feature selection mask, $q_i^{(b)} = 1$ if $q_i \geq T$, 0 otherwise, where T is a predefined threshold. The resulting patient-specific fingerprint $\mathbf{f}^s$ is therefore obtained: $\mathbf{f}^s = \mathbf{f} \odot \mathbf{q}^{(b)}$, where $\mathbf{q}^{(b)} = [q_1^{(b)}, q_2^{(b)}, ..., q_{3JK}^{(b)}]^\top$. This sparse fingerprint is then used for final prediction on new test data: $\hat{\mathbf{c}} = R_\psi(\mathbf{f}^s)$.

3   Experiments and Results

3.1   Dataset and Implementation Details

Dataset and Preprocessing: Our methodology is demonstrated on the MRNet dataset [3], comprising 1,370 3D knee MRI scans collected across axial, coronal, and sagittal planes. Each exam is annotated for three diagnostic labels: general abnormalities, ACL tear, and meniscal tear. We use 1,130 studies for development set and 120 for test. All images are preprocessed by resizing them to $32 \times 128 \times 128$ voxels. To ensure consistent ROI definition and patch division, we apply affine registration using NiftyReg [10], and evaluate performance with and without this step. The ROI is defined as a subvolume encompassing 50% of the original image depth, 30% of the height, and 50% of the width, which is empirically determined by radiologists to effectively cover anatomical regions commonly associated with ACL and meniscus pathology.

Table 1. Comparison with end-to-end DL models. N: Number of patches; PFS: Patient-specific features selection; Reg: Registration in preprocessing. * denotes reproduced results

Method	N	PFS	View	Reg	Type	Evaluation Metrics			
						Acc	Sen	Spe	AUC
Ours	2*2*2	✓	3	✓	abn	**0.92 ± 0.09**	0.98 ± 0.03	0.71 ± 0.33	0.85 ± 0.17
					acl	**0.94 ± 0.09**	0.96 ± 0.06	0.88 ± 0.05	0.92 ± 0.04
					men	**0.84 ± 0.16**	0.97 ± 0.04	0.67 ± 0.34	0.82 ± 0.18
MRNet*	–	–	3	–	abn	0.85 ± 0.01	0.84 ± 0.03	0.65 ± 0.09	0.94 ± 0.02
					acl	0.86 ± 0.02	0.77 ± 0.02	0.97 ± 0.02	0.97 ± 0.02
					men	0.71 ± 0.04	0.69 ± 0.03	0.74 ± 0.03	0.84 ± 0.04
ELNet*	–	–	3	–	abn	0.80 ± 0.01	0.95 ± 0.01	0.22 ± 0.03	0.73 ± 0.01
					acl	0.70 ± 0.10	0.72 ± 0.26	0.68 ± 0.04	0.73 ± 0.08
					men	0.65 ± 0.06	0.61 ± 0.13	0.63 ± 0.07	0.69 ± 0.03
SKID	–	–	3	–	abn	0.87	0.98	0.49	0.90
					acl	0.80	0.74	0.85	0.89
					men	0.73	0.92	0.57	0.81

Feature Extraction And Dynamic Selection: Radiomic features are extracted using `PyTorchRadiomics`, a PyTorch-based implementation built upon the PyRadiomics library [13]. We compute a comprehensive set of features, including *First-Order Statistics* (19 features), *3D Shape Descriptors* (16 features), and texture-based features such as *Gray Level Co-occurrence Matrix (GLCM)* (24 features), *Gray Level Run Length Matrix (GLRLM)* (16 features), *Gray Level Size Zone Matrix (GLSZM)* (16 features), *Neighbouring Gray Tone Difference Matrix* (5 features), and *Gray Level Dependence Matrix (GLDM)* (14 features). To capture localized tissue characteristics, we apply spatial decomposition by dividing each pathological ROI into patches at multiple granularities: a single patch ($1 \times 1 \times 1$, i.e., using the entire ROI as input), 8 patches ($2 \times 2 \times 2$), and 27 patches ($3 \times 3 \times 3$). For each configuration, features are independently extracted across axial, coronal, and sagittal view, resulting in a high-dimensional radiomic feature pool that comprehensively represents 3D tissue characteristics. A 3DResNet-18 [6] serves as the feature-weighting neural network, learning patient-specific relevance scores to adaptively select discriminative features for individual cases.

Classification Model: As described in Sect. 2, the weighted feature vector $\mathbf{f}^{w}$ is passed to a classifier R_{ψ} to predict class probabilities. In this work, we implement R_{ψ} as an extended logistic regression model that incorporates both first-order individual feature contributions and second-order pairwise interactions. This design enables the classifier to capture second-order dependencies among selected features, should they be predictive for the classification, while maintaining relatively high explainability. During model development, values of the task-specific classification thresholds T are determined based on the best Youden's index obtainable on the validation set.

Comparison and Ablation Experiments: We compared our method with established end-to-end DL models, including MRNet [3], ELNet [12], and SKID [9]. To analyze the impact of core components, we conducted ablation studies focusing on three aspects: radiomic feature composition, feature selection mechanism, and spatial decomposition. Using the full radiomic feature set consistently outperformed configurations restricted to second-order or first-order & shape features. Disabling the feature-weighting network and using unweighted features reduced accuracy, confirming the value of adaptive selection. Finally, among various patch sizes, the $2 \times 2 \times 2$ configuration delivered the most favorable balance between diagnostic performance and model explainability (specific examples are discussed in Sect. 3.2).

3.2   Results and Interpretation

Quantitative Evaluation: For ACL tear classification, our patient-specific radiomic signature framework achieves an accuracy of 0.94 ± 0.09, AUC of 0.92 ± 0.04, and sensitivity of 0.96 ± 0.06. Compared to MRNet and ELNet, our method demonstrates consistently strong performance. Notably, statistically significant improvements were observed in specificity (vs. ELNet, $p = 0.0002$) and in accuracy (vs. ELNet, $p = 0.012$). For meniscus tear classification, our method also achieves superior performance, particularly in accuracy ($p = 0.028$ vs. ELNet), highlighting its robustness across multiple diagnostic tasks.

Table 2. Ablation study evaluating different configuration settings. N: Number of patches; RF: Radiomic features PFS: Patient-specific features selection; Reg: Registration in preprocessing.

Method	N	RF Type	FS	View	Reg	Type	Evaluation Metrics			
							Acc	Sen	Spe	AUC
Ours	2*2*2	All	✓	3	✓	abn	**0.92 ± 0.09**	0.98 ± 0.03	0.71 ± 0.33	0.85 ± 0.17
						acl	**0.94 ± 0.09**	0.96 ± 0.06	0.88 ± 0.05	0.92 ± 0.04
						men	0.84 ± 0.16	0.97 ± 0.04	0.67 ± 0.34	0.82 ± 0.18
Ours	3*3*3	All	✓	3	✓	abn	0.91 ± 0.11	0.98 ± 0.08	0.65 ± 0.41	0.83 ± 0.21
						acl	0.94 ± 0.11	0.89 ± 0.20	0.98 ± 0.04	0.94 ± 0.12
						men	0.83 ± 0.16	0.74 ± 0.31	0.89 ± 0.22	0.81 ± 0.16
2nd RF	2*2*2	2nd	✓	3	✓	abn	0.90 ± 0.08	0.98 ± 0.03	0.68 ± 0.25	0.83 ± 0.14
						acl	0.83 ± 0.17	0.91 ± 0.12	0.74 ± 0.26	0.82 ± 0.18
						men	**0.88 ± 0.14**	0.92 ± 0.13	0.85 ± 0.16	0.88 ± 0.13
1st RF	2*2*2	1st+shape	✓	3	✓	abn	0.77 ± 0.13	0.75 ± 0.19	0.87 ± 0.18	0.81 ± 0.10
						acl	0.82 ± 0.14	0.79 ± 0.19	0.83 ± 0.13	0.81 ± 0.14
						men	0.72 ± 0.10	0.80 ± 0.15	0.68 ± 0.09	0.74 ± 0.10
1patch	1*1*1	All	✓	3	✓	abn	0.73 ± 0.32	0.66 ± 0.44	0.93 ± 0.13	0.79 ± 0.21
						acl	0.76 ± 0.14	0.97 ± 0.05	0.58 ± 0.24	0.77 ± 0.12
						men	0.62 ± 0.07	0.78 ± 0.25	0.55 ± 0.27	0.66 ± 0.04
NoFS	2*2*2	All	✗	3	✓	abn	0.88 ± 0.12	0.89 ± 0.14	0.90 ± 0.22	0.89 ± 0.13
						acl	0.92 ± 0.04	0.96 ± 0.05	0.93 ± 0.12	0.94 ± 0.08
						men	0.83 ± 0.17	0.90 ± 0.07	0.81 ± 0.24	0.85 ± 0.14
NoReg	2*2*2	All	✓	3	✗	abn	0.77 ± 0.08	0.59 ± 0.13	0.95 ± 0.04	0.75 ± 0.07
						acl	0.71 ± 0.09	0.43 ± 0.18	0.98 ± 0.04	0.68 ± 0.10
						men	0.65 ± 0.14	0.97 ± 0.04	0.43 ± 0.26	0.69 ± 0.13

Table 2 also presents key ablation results. When evaluating radiomic feature types, the full feature set (first-order, shape, and texture features) consistently yields the highest performance. Limiting the model to only second-order features slightly degrades overall accuracy, while using only first-order and shape features results in notable performance drop. Removing the feature-weighting neural network and using unweighted features (NoFS) leads to a noticeable decline in performance, confirming the critical role of adaptive feature selection.

Feature Analysis and Case Study: Based on feature analysis of the ACL tear test set, the most relevant radiomic features primarily originate from the sagittal view, which aligns with clinical practice, where ACL tears are most clearly visualized. When predicting the presence of an ACL tear, key features tend to concentrate in patches that correspond anatomically to the ACL location. However, in some cases, key features appear dispersed, reflecting secondary structural changes—such as bone marrow edema and anterior tibial translation— that extend beyond the ligament itself and affect the surrounding tissues. In cases of ACL tear, the top-ranked features are predominantly first-order statistics and

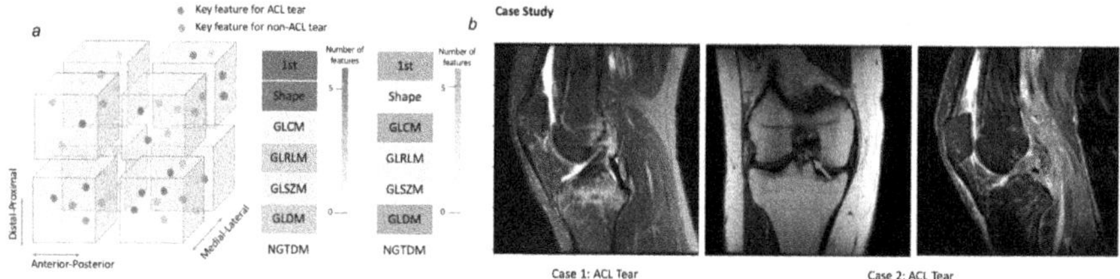

Fig. 2. (a) Anatomical location and feature type distribution of the top 20 top radiomic features; (b) representative cases of ACL tears (details in Sect.3.2).

shape-based features, with additional contribution from texture features. These features effectively capture macrostructural abnormalities, including ligament discontinuity, joint effusion, and altered signal intensity. In contrast, non-ACL tear cases appear to rely more on texture features, particularly *GLCM* and *GLDM*. This suggests that, when there is no obvious structural disruption, texture analysis may provide valuable information about subtle tissue heterogeneity.

We further analyse two representative cases with differing severities of ACL injury. Case 1 shows residual fibres attached to the tibial insertion, while Case 2 demonstrates loss of normal fibre architecture (as indicated by red arrows), accompanied by more extensive bone marrow edema and joint effusion. In both cases, the most important radiomic features include Large Area Low Gray Level Emphasis, Gray Level Variance, and Sum Entropy. These features reflect, respectively, the extent of low-intensity regions associated with joint fluid, variability in signal due to disrupted fibre structure, and increased complexity of signal distribution within the lesion zone. The most notable feature between the two cases is Large Dependence Low Gray Level Emphasis (LDLGLE), which quantifies large, homogeneous low-intensity areas. The case with more widespread disruption shows higher LDLGLE values, consistent with a greater extent of uniform signal alteration. This distinction approximates radiological interpretation, which is influenced by differences in the extent of ligamentous injury.

4   Conclusion and Discussion

We propose a DL–guided, patient-specific radiomic framework for knee MRI diagnosis that integrates classical handcrafted features with modern data-driven modeling. By dynamically selecting relevant features tailored to individual imaging characteristics and applying neural weighting, the method leverages the interpretability of traditional radiomics while benefiting from the representational power of DL. The final transparent logistic regression classifier ensures interpretability, offering clinicians actionable insights into which features drive each prediction—an essential quality often lacking in DL–based medical imaging models.

While our framework improves interpretability, radiomic features may still be sensitive to variations in imaging protocols, such as scanner type and sequence settings. These differences can introduce non-biological variability, affecting generalizability. To address this, we incorporated image registration to reduce anatomical differences across scans. Nevertheless, additional harmonization strategies—such as domain adaptation—may be necessary to ensure further robustness across diverse imaging environments.

Acknowledgement. This work was supported by the International Alliance for Cancer Early Detection, an alliance between Cancer Research UK [EDDAPA-2024/100014] & [C73666/A31378], Canary Center at Stanford University, the University of Cambridge, OHSU Knight Cancer Institute, University College London and the University of Manchester; and the National Institute for Health Research University College London Hospitals Biomedical Research Centre.

References

1. Al Mohammad, B., Gharaibeh, M.A.: Magnetic resonance imaging of anterior cruciate ligament injury. Orthopedic Res. Rev., 233–242 (2024)
2. Ardakani, A.A., Bureau, N.J., Ciaccio, E.J., Acharya, U.R.: Interpretation of radiomics features-a pictorial review. Comput. Methods Programs Biomed. **215**, 106609 (2022)
3. Bien, N., et al.: Deep-learning-assisted diagnosis for knee magnetic resonance imaging: development and retrospective validation of mrnet. PLoS Med. **15**(11), e1002699 (2018)
4. Chen, H., Gomez, C., Huang, C.M., Unberath, M.: Explainable medical imaging ai needs human-centered design: guidelines and evidence from a systematic review. NPJ Digital Med. **5**(1), 156 (2022)
5. Chen, Y., Ni, S., Ivanova, A., Saeed, S.U., Hargunani, R., Huang, J., Liu, C., Hu, Y.: Patient-specific radiomic feature selection with reconstructed healthy persona of knee mr images. arXiv preprint arXiv:2503.13131 (2025)
6. He, K., Zhang, X., Ren, S., Sun, J.: Deep residual learning for image recognition. In: Proceedings of the IEEE Conference on Computer Vision and Pattern Recognition, pp. 770–778 (2016)
7. Kobayashi, K., Miyake, M., Takahashi, M., Hamamoto, R.: Observing deep radiomics for the classification of glioma grades. Sci. Rep. **11**(1), 10942 (2021)
8. Lavrova, E., Woodruff, H.C., Khan, H., Salmon, E., Lambin, P., Phillips, C.: A review of handcrafted and deep radiomics in neurological diseases: transitioning from oncology to clinical neuroimaging. arXiv preprint arXiv:2407.13813 (2024)
9. Manna, S., Bhattacharya, S., Pal, U.: Self-supervised representation learning for knee injury diagnosis from magnetic resonance data. IEEE Trans. Artifi. Intell. (2023)
10. Modat, M., Ridgway, G.R., Taylor, Z.A., Lehmann, M., Barnes, J., Hawkes, D.J., et al.: Fast free-form deformation using graphics processing units. Comput. Methods Programs Biomed. **98**(3), 278–284 (2010)
11. Salzler, M., et al.: State-of-the-art anterior cruciate ligament tears: a primer for primary care physicians. Phys, Sportsmedicine **43**(2), 169–177 (2015)
12. Tsai, C.H., Kiryati, N., Konen, E., Eshed, I., Mayer, A.: Knee injury detection using mri with efficiently-layered network (elnet). In: Medical Imaging with Deep Learning, pp. 784–794. PMLR (2020)

13. Van Griethuysen, J.J., et al.: Computational radiomics system to decode the radiographic phenotype. Can. Res. **77**(21), e104–e107 (2017)
14. Zhang, W., Guo, Y., Jin, Q.: Radiomics and its feature selection: a review. Symmetry **15**(10), 1834 (2023)

Prototype-Enhanced Confidence Modeling for Cross-Modal Medical Image-Report Retrieval

Shreyank N. Gowda[1]([✉]), Xiaobo Jin[2], and Christian Wagner[1]

[1] School of Computer Science, The University of Nottingham, NG8 1BB Nottingham, UK
shreyank.narayanagowda@nottingham.ac.uk

[2] Department of Intelligent Science, Xi'an Jiaotong-Liverpool University, Suzhou 215123, China

Abstract. In cross-modal retrieval tasks, such as image-to-report and report-to-image retrieval, accurately aligning medical images with relevant text reports is essential but challenging due to the inherent ambiguity and variability in medical data. Existing models often struggle to capture the nuanced, multi-level semantic relationships in radiology data, leading to unreliable retrieval results. To address these issues, we propose the Prototype-Enhanced Confidence Modeling (PECM) framework, which introduces multi-level prototypes for each modality to better capture semantic variability and enhance retrieval robustness. PECM employs a dual-stream confidence estimation that leverages prototype similarity distributions and an adaptive weighting mechanism to control the impact of high-uncertainty data on retrieval rankings. Applied to radiology image-report datasets, our method achieves significant improvements in retrieval precision and consistency, effectively handling data ambiguity and advancing reliability in complex clinical scenarios. We report results on multiple different datasets and tasks including fully supervised and zero-shot retrieval obtaining performance gains of up to 10.17%, establishing in new state-of-the-art.

Keywords: Prototype · Uncertainty · Retrieval · Vision-Language

1 Introduction

Modern healthcare generates vast digital medical records including radiology images, creating both opportunity and challenge: effectively retrieving relevant cross-modal information is crucial for improving clinical decision-making. Cross-modal retrieval aligns and retrieves related data across modalities–using a query image to find relevant text or vice versa. Adapting cross-modal retrieval to clinical domains introduces unique obstacles. Radiology images and reports contain specialized content with inherent ambiguities: overlapping visual features across conditions [14] and limited descriptive granularity in reports [5]. Recent

© The Author(s), under exclusive license to Springer Nature Switzerland AG 2026

M. Reyes et al. (Eds.): iMIMIC 2025, LNCS 16464, pp. 75–84, 2026.
https://doi.org/10.1007/978-3-032-17611-0_8

deep learning advances have improved retrieval capabilities through attention mechanisms and contrastive learning. MCR [25] creates robust image-report embeddings for medical retrieval tasks. Vision-Language Pre-training models like CLIP [20] and its medical adaptations [27] enhance performance by transferring knowledge to downstream medical tasks. Other works [3,5,26,28] further align visual and linguistic medical representations. Most existing methods assume consistent data quality [3,25,26], rarely true in medical imaging. Prototype-based methods [12] capture diverse semantic structures by learning representative prototypes for each modality. Li et al. [12] quantify aleatoric uncertainty in cross-modal retrieval, improving reliability by accounting for data quality variations.

We propose a novel *Prototype-Enhanced Confidence Modeling (PECM)* framework to improve robustness of medical cross-modal retrieval by quantifying representation uncertainty. Unlike traditional models, PECM explicitly considers ambiguity in radiology data through multi-level prototypes that capture different semantic aspects within images and reports. This enables representation of subtle differences across cases, improving retrieval accuracy where conventional models might be misled by ambiguous features. While standard cross-modal retrieval embeds images and text into a shared space using cosine similarity, PECM introduces a *dual-stream confidence estimation* that models uncertainty by assessing dissimilarity within prototype similarity distributions and employs an *adaptive weighting mechanism*. This adjusts the influence of high-uncertainty data points, balancing accuracy and reliability without disregarding informative cases. The resulting re-ranking system prioritizes high-confidence matches while still considering lower-confidence data.

Our key contributions are:

1. Prototype-Enhanced Representation: Multi-level prototypes represent unique semantic categories within each modality, providing finer-grained understanding of radiology data by capturing variations at different semantic levels, enhancing generalization across diverse medical datasets.

2. Dual-Stream Confidence Estimation: Using dissimilarity within prototype distributions as an uncertainty proxy, our model adaptively weights uncertain cases. This achieves informed balancing of information, effectively balancing accuracy and informativeness without discarding useful data.

3. Adaptive Re-ranking Mechanism: We implement adaptive ranking that leverages confidence information to dynamically adjust retrieval order, prioritizing high-confidence matches. This enriches retrieval by integrating uncertainty as an informative signal, enhancing robustness.

2   Method

This section introduces the Prototype-Enhanced Confidence Modeling (PECM) framework for cross-modal retrieval, designed to quantify uncertainty in medical imaging data. We show an overview in Fig. 1.

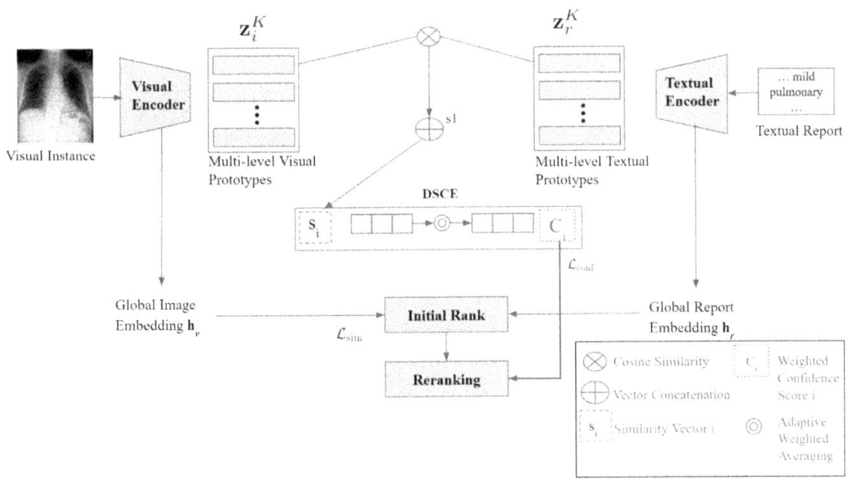

Fig. 1. PECM processes paired image-report data through encoders to generate multi-level prototypes. The Dual-Stream Confidence Estimation calculates similarities between corresponding prototypes, weighted to produce confidence scores. Initial ranking uses global similarity, while re-ranking incorporates confidence-weighted similarity. The model is trained with a combination of similarity loss ($\mathcal{L}_{\mathrm{sim}}$), confidence loss ($\mathcal{L}_{\mathrm{conf}}$), and diversity loss ($\mathcal{L}_{\mathrm{diversity}}$) (not illustrated here for simplicity).

2.1 Multi-level Prototype Construction

To address variability in medical data, we create multi-level prototypes for each modality to capture diverse semantic meanings. With K prototypes per modality, each represents a distinct level of semantic granularity.

For images, our visual encoder (ViT-B) divides 224×224 images into 16×16 patches, producing 196 patches (14×14 grid). To reduce computation, we group these into 3×3 grids, creating 25 regional prototypes across the image (5×5 regions). We add a global prototype from the [CLS] token, resulting in $K = 26$ prototypes per image (Fig. 2). For text, we divide each report into $K-1$ sentence groups and process each through a paragraph embedding model [11] to obtain region-level prototypes. These combine with the global document prototype to form K prototypes per report. We denote image and report prototypes as $\mathbf{Z}_i = \{\mathbf{z}_i^1, \mathbf{z}_i^2, \ldots, \mathbf{z}_i^K\}$ and $\mathbf{Z}_r = \{\mathbf{z}_r^1, \mathbf{z}_r^2, \ldots, \mathbf{z}_r^K\}$, respectively.

2.2 Dual-Stream Confidence Estimation

Our dual-stream confidence estimation quantifies image-text match reliability by aligning corresponding prototypes. Image and report modalities are represented as prototype sets $\mathbf{Z}_i = \{\mathbf{z}_i^1, \mathbf{z}_i^2, \ldots, \mathbf{z}_i^K\}$ and $\mathbf{Z}_r = \{\mathbf{z}_r^1, \mathbf{z}_r^2, \ldots, \mathbf{z}_r^K\}$. For each prototype pair $(\mathbf{z}_i^k, \mathbf{z}_r^k)$, we compute cosine similarity:

$$\mathrm{sim}(\mathbf{z}_i^k, \mathbf{z}_r^k) = \frac{\mathbf{z}_i^k \cdot \mathbf{z}_r^k}{\|\mathbf{z}_i^k\|\|\mathbf{z}_r^k\|} \tag{1}$$

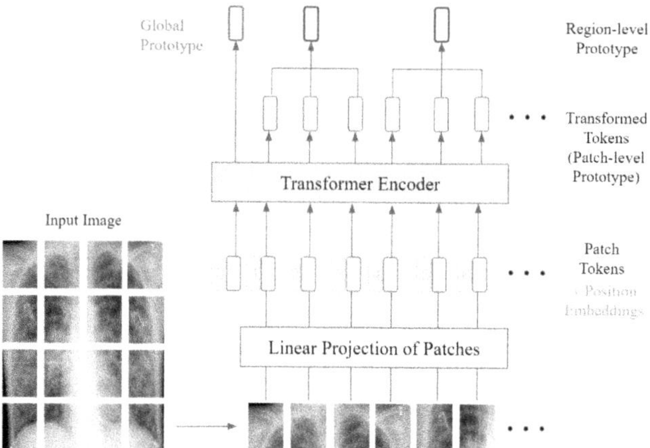

Fig. 2. Multi-Level Prototype Construction. Illustration of prototype calculation: images are patched, processed through transformer encoding, averaged into patch-level prototypes, grouped into regional prototypes, and combined with global prototype to form multi-level representation. 16 patches for illustration only.

These similarities form vector $\mathbf{s} = [\mathrm{sim}(\mathbf{z}_i^1, \mathbf{z}_r^1), \mathrm{sim}(\mathbf{z}_i^2, \mathbf{z}_r^2), \ldots, \mathrm{sim}(\mathbf{z}_i^K, \mathbf{z}_r^K)]$. The final confidence score c is their weighted average:

$$c = \frac{1}{K} \sum_{k=1}^{K} \mathrm{sim}(\mathbf{z}_i^k, \mathbf{z}_r^k) \cdot w_k \tag{2}$$

where adaptive weights w_k emphasize high-confidence pairs and reduce influence of ambiguous matches.

2.3 Adaptive Re-ranking Mechanism

PECM initially ranks retrieval based on global similarity between image and report representations:

$$\mathrm{sim}_{\mathrm{initial}}(v, t) = \frac{\mathbf{h}_v \cdot \mathbf{h}_t}{\|\mathbf{h}_v\| \|\mathbf{h}_t\|} \tag{3}$$

where global embeddings $\mathbf{h}_v$ and $\mathbf{h}_t$ are weighted averages of their respective prototypes:

$$\mathbf{h}_v = \sum_{k=1}^{K} w_k \mathbf{z}_i^k, \quad \mathbf{h}_t = \sum_{k=1}^{K} w_k \mathbf{z}_r^k \tag{4}$$

with shared weights w_k for cross-modal alignment. For query image v and candidate reports $\{t_j\}_{j=1}^{N}$, initial ranking is:

$$\mathrm{rank}_{\mathrm{initial}}(v, t_j) = \mathrm{Rank}_{\mathrm{desc}}\left(\mathrm{sim}_{\mathrm{initial}}(v, t_j) : j = 1, \ldots, N\right) \tag{5}$$

We then calculate confidence scores for adaptive re-ranking:

$$C(v, t_j) = \frac{1}{K} \sum_{k=1}^{K} \mathrm{sim}(\mathbf{z}_i^k, \mathbf{z}_r^k) \cdot w_k \tag{6}$$

The final score weights initial similarity with confidence:

$$R(v, t_j) = \mathrm{sim}_{\mathrm{initial}}(v, t_j) \cdot C(v, t_j) \tag{7}$$

Final ranking prioritizes high-confidence matches:

$$\mathrm{rank}_{\mathrm{final}}(v, t_j) = \mathrm{Rank}_{\mathrm{desc}}\left(R(v, t_j) : j = 1, \ldots, N\right) \tag{8}$$

2.4 Objective Function

PECM training uses a triple-component objective function:

$$\mathcal{L} = \mathcal{L}_{\mathrm{sim}} + \lambda \mathcal{L}_{\mathrm{conf}} + \mu \mathcal{L}_{\mathrm{div}} \tag{9}$$

where λ and μ are balancing hyperparameters. The similarity loss uses contrastive learning to promote match quality:

$$\mathcal{L}_{\mathrm{sim}} = -\log \frac{\exp(\mathrm{sim}_{\mathrm{initial}}(v, t))}{\sum_{j=1}^{N} \exp(\mathrm{sim}_{\mathrm{initial}}(v, t_j))} \tag{10}$$

The confidence loss optimizes prototype weights, penalizing low-confidence matches:

$$\mathcal{L}_{\mathrm{conf}} = \frac{1}{N} \sum_{i=1}^{N} \left(1 - C(v_i, t_i)\right)^2 \tag{11}$$

where $C(v_i, t_i)$ represents weighted similarity between corresponding prototypes. The diversity loss ensures prototypes within each modality remain distinct:

$$\mathcal{L}_{\mathrm{div}} = \sum_{k=1}^{K} \sum_{l \neq k}^{K} \left(1 - \mathrm{sim}(\mathbf{z}^k, \mathbf{z}^l)\right)^2 \tag{12}$$

where $\mathbf{z}^k$ and $\mathbf{z}^l$ are prototypes within the same modality, encouraging each to capture unique semantic features.

This objective balances similarity, confidence, and diversity for accurate and reliable retrieval across ambiguous medical data. We set μ and λ as 1 based on experimental results.

3   Experimental Analysis

3.1   Datasets

Following MCR [25], we use MIMIC-CXR [9] (377,100 images, 65,379 patients, 227,835 reports) for cross-modal retrieval, extracting 'findings' sections and excluding incomplete samples. For generalizable retrieval, we combine RadImageNet [17], NIH14 [22], MIMIC [9], and CheXpert [8] per [2], creating a diverse dataset (1.6M+ images, 4 modalities, 12 anatomical regions, 185 classes) with duplicate removal, label standardization, and single-label focus. Additional evaluations use ROCO from ImageCLEF 2023 [7] (60,918/10,437/10,473 train/val/test images with captions and CUIs) and MURA [21] (40,895 musculoskeletal radiographs across seven body parts).

3.2   Comparison to State-of-the-Art

Table 1. The performance comparison of different cross-modal retrieval methods on the MIMIC-CXR dataset. VLP and MVLP represent Vision-Language Pre-training and Medical Vision-Language Pre-training, respectively. MMU and CMA represent Multi-Modal Fusion and Cross-Modal Alignment, respectively.

Method	Category	I → R			R → I		
		Recall@1	Recall@5	Recall@10	Recall@1	Recall@5	Recall@10
ConVIRT [28]	MVLP, CMA	6.765%	22.421%	32.219%	6.647%	16.151%	23.832%
REFERS [29]	MVLP, CMA	8.294%	25.765%	36.910%	8.607%	18.929%	28.755%
LoVT [18]	MVLP, CMA	10.601%	30.715%	42.017%	11.504%	24.855%	36.067%
DiVE [18]	CMA	12.960%	35.070%	48.419%	13.294%	26.993%	38.780%
CLIP [20]	VLP, CMA	14.256%	36.392%	48.263%	14.700%	29.342%	40.160%
BLIP2 [13]	VLP, CMA	17.159%	41.498%	53.447%	18.875%	34.974%	47.657%
M3AE [1]	MVLP, MMU	15.267%	42.898%	56.739%	17.171%	36.153%	50.135%
MaskCLIP [3]	MVLP, CMA	18.170%	43.339%	55.962%	19.301%	36.531%	49.058%
CXR-CLIP [26]	MVLP, CMA	20.736%	45.853%	58.968%	22.985%	40.477%	52.484%
MCR + MbA [25]	MVLP, CMA	24.598%	52.281%	65.241%	27.354%	45.709%	58.758%
SENSE [15]	MVLP, CMA	19.500%	45.100%	57.300%	19.800%	46.200%	57.500%
PECM (Ours)	MVLP, CMA	**28.871%**	**58.644%**	**69.691%**	**32.431%**	**50.946%**	**64.252%**

Cross-Modal Retrieval. For cross-modal retrieval, we evaluate image-to-report (I→R) and report-to-image (R→I) tasks. Following MCR [25], we use Recall@K (K ∈ {1,5,10}) as our metric, measuring the percentage of correct matches within the top K retrievals. For R→I, recall is capped by min(K, number of images linked to the report). Table 1 shows our approach achieves performance gains up to 6.363% on MIMIC-CXR.

Table 2. Content-Based Image Retrieval performance of the evaluated models on the combined dataset.

Method	Sample-wise (micro)				Class-wise (macro)			
	P@1	P@3	P@5	P@10	P@1	P@3	P@5	P@10
ViT [4]	0.560	0.554	0.550	0.543	0.217	0.204	0.199	0.190
SAM [10]	0.520	0.517	0.515	0.511	0.197	0.187	0.182	0.175
MedSAM [16]	0.489	0.483	0.478	0.470	0.154	0.146	0.142	0.134
CLIP [20]	0.571	0.567	0.565	0.559	0.222	0.213	0.209	0.202
MedCLIP [23]	0.566	0.561	0.559	0.554	0.223	0.211	0.205	0.198
BiomedCLIP [27]	0.594	0.590	0.588	0.583	0.240	0.230	0.224	0.217
M3AE [1]	0.513	0.507	0.503	0.497	0.183	0.177	0.172	0.166
DINOv2 [19]	0.553	0.548	0.545	0.540	0.219	0.204	0.199	0.192
X-MIR [6]	0.478	0.471	0.466	0.457	0.149	0.142	0.138	0.131
KL-CVR [24]	0.602	0.596	0.593	0.585	0.245	0.234	0.229	0.211
PECM (Ours)	**0.631**	**0.627**	**0.616**	**0.605**	**0.257**	**0.249**	**0.241**	**0.219**

Generalizable Retrieval. Using our pre-trained visual encoder, we compare against ViT [4], SAM [10], MedSAM [16], CLIP [20], MedCLIP [23], Biomed-CLIP [27], M3AE [1], DINOv2 [19], X-MIR [6], and KL-CVR [24]. We report both micro- and macro-averaged scores to ensure fair evaluation across imbalanced datasets. Table 2 shows up to 2.9% improvement.

Zero-Shot Retrieval. We compare PECM against state-of-the-art methods on ROCO and MURA without fine-tuning. Our model, pre-trained across various modalities and anatomical regions, outperforms methods specifically fine-tuned on these benchmarks. While further fine-tuning yields slight improvements, PECM achieves competitive or superior performance in zero-shot settings, surpassing current state-of-the-art [24]. This demonstrates PECM's robust generalization across diverse medical datasets, with performance gains up to 10.17% as shown in Table 3.

3.3   Ablation Study

We assess each component's contribution through ablation studies of Prototype-Enhanced Representation (PER), Dual-Stream Confidence Estimation (DSCE), and Adaptive Re-ranking (ARR). Table 4 shows Recall@5 for I→R and R→I tasks. The baseline without these components performs poorest. Adding PER improves performance by capturing semantic nuances across modalities. DSCE further boosts recall through enhanced match reliability. The full PECM framework with all components achieves the best results, demonstrating their complementary effects. We also ablate the different loss functions in Table 5.

Table 3. Generalizing CIBR to other datasets

Methods	ROCO CUI@K			MURA P@K		
	@5	@10	@50	@5	@10	@30
General initialization methods						
ImageNet	35.53	37.39	39.18	84.42	77.23	70.04
CLIP [20]	35.72	37.39	39.20	65.70	54.38	46.55
In-domain initialization methods: Fully supervised						
CLIP [20]	39.85	40.84	44.86	69.96	57.82	50.55
BioMed CLIP [27]	42.49	45.07	**47.23**	74.34	65.89	56.50
KL-CVR [24]	40.68	43.44	46.46	**77.83**	68.56	60.80
In-domain initialization methods: Zero-Shot						
KL-CVR [24]	33.67	35.51	38.72	66.52	64.65	55.89
PECM (Ours)	**42.95**	**45.23**	47.23	76.69	**68.59**	**60.92**

Table 4. Ablation study of the proposed components. PER = Prototype-Enhanced Representation, DSCE = Dual-Stream Confidence Estimation, ARR = Adaptive Re-ranking. Reported results are for Recall @ 5 on MIMIC-CXR.

PER	DSCE	ARR	I → R	R → I
✗	✗	✗	36.572%	29.781%
✓	✗	✗	44.392%	37.751%
✗	✓	✗	41.922%	36.345%
✗	✗	✓	39.472%	33.915%
✓	✓	✗	54.452%	45.278%
✓	✗	✓	50.794%	45.046%
✗	✓	✓	52.304%	47.116%
✓	✓	✓	**58.644%**	**50.946%**

Table 5. Ablation study of individual loss components. Performance is reported as Recall @ 5 for image-to-report (I→R) and report-to-image (R→I) tasks. Results on MIMIC-CXR Dataset.

$\mathcal{L}_{sim}$	$\mathcal{L}_{conf}$	$\mathcal{L}_{div}$	I→R	R→I
✓	✗	✗	55.2	46.8
✗	✓	✗	50.1	43.5
✗	✗	✓	48.5	42.0
✓	✓	✗	57.1	49.0
✓	✗	✓	56.8	48.1
✗	✓	✓	52.7	44.8
✓	✓	✓	**58.6**	**50.9**

3.4 Implementation Details

PECM is implemented in PyTorch using two pre-trained encoders: ViT-B (visual) and BERT (text). For Prototype-Enhanced Representation, images are divided into 16 patches, processed through ViT, and grouped into 3×3 regions to create 25 regional prototypes plus one global [CLS] prototype. Text prototypes are generated from doc2vec paragraph embeddings. DSCE computes prototype similarities with adaptive weights learned during training. Initial ranking uses global similarity, while Adaptive Re-ranking incorporates confidence scores to prioritize reliable matches. Training uses image-report pairs from multiple medical datasets with three loss components (contrastive, confidence, diversity) for 30 epochs (batch size 32, Adam optimizer, cosine annealing from 1e-4).

4   Conclusion

In this work, we addressed cross-modal retrieval challenges in medical imaging by developing the Prototype-Enhanced Confidence Modeling (PECM) framework. PECM introduces multi-level prototypes for both image and text modalities, capturing fine-grained semantic layers like anatomical and pathological details. Our Dual-Stream Confidence Estimation quantifies prototype alignment reliability, while Adaptive Re-ranking prioritizes confident matches. Experiments demonstrate PECM's superior performance across multiple medical retrieval tasks. By effectively handling both semantic complexity and uncertainty, PECM advances the state-of-the-art in medical cross-modal retrieval, providing a robust solution with clear potential for clinical applications.

References

1. Chen, Z., et al.: Multi-modal masked autoencoders for medical vision-and-language pre-training. In: International Conference on Medical Image Computing and Computer-Assisted Intervention, pp. 679–689. Springer (2022). https://doi.org/10.1007/978-3-031-16443-9_65
2. Denner, S., et al.: Leveraging foundation models for content-based medical image retrieval in radiology. arXiv preprint arXiv:2403.06567 (2024)
3. Dong, X., et al.: Maskclip: masked self-distillation advances contrastive language-image pretraining. In: Proceedings of the IEEE/CVF Conference on Computer Vision and Pattern Recognition, pp. 10995–11005 (2023)
4. Dosovitskiy, A., et al.: An image is worth 16x16 words, vol. 7. arXiv preprint arXiv:2010.11929 (2020)
5. Gowda, S.N., Clifton, D.A.: Masks and manuscripts: Advancing medical pre-training with end-to-end masking and narrative structuring. In: International Conference on Medical Image Computing and Computer-Assisted Intervention, pp. 426–436. Springer (2024). https://doi.org/10.1007/978-3-031-72120-5_40
6. Hu, B., Vasu, B., Hoogs, A.: X-mir: explainable medical image retrieval. In: Proceedings of the IEEE/CVF Winter Conference on Applications of Computer Vision, pp. 440–450 (2022)
7. Ionescu, B., et al.: Overview of the imageclef 2023: multimedia retrieval in medical, social media and internet applications. In: International Conference of the Cross-Language Evaluation Forum for European Languages. pp. 370–396. Springer (2023). https://doi.org/10.1007/978-3-031-42448-9_25
8. Irvin, J., et al.: Chexpert: a large chest radiograph dataset with uncertainty labels and expert comparison. In: Proceedings of the AAAI Conference on Artificial Intelligence, vol. 33, pp. 590–597 (2019)
9. Johnson, A.E., et al.: Mimic-cxr, a de-identified publicly available database of chest radiographs with free-text reports. Sci. Data **6**(1), 317 (2019)
10. Kirillov, A., et al.: Segment anything. In: Proceedings of the IEEE/CVF International Conference on Computer Vision, pp. 4015–4026 (2023)
11. Le, Q., Mikolov, T.: Distributed representations of sentences and documents. In: International Conference on Machine Learning, pp. 1188–1196. PMLR (2014)
12. Li, H., Song, J., Gao, L., Zhu, X., Shen, H.: Prototype-based aleatoric uncertainty quantification for cross-modal retrieval. Adv, Neural Inform. Process. Syst. **36** (2024)

13. Li, J., Li, D., Savarese, S., Hoi, S.: Blip-2: bootstrapping language-image pre-training with frozen image encoders and large language models. In: International Conference on Machine Learning, pp. 19730–19742. PMLR (2023)
14. Li, X., Cao, R., Zhu, D.: Vispi: automatic visual perception and interpretation of chest x-rays. In: Medical Imaging with Deep Learning (2020)
15. Liu, B., Lu, Z., Wang, Y.: Towards medical vision-language contrastive pre-training via study-oriented semantic exploration. In: Proceedings of the 32nd ACM International Conference on Multimedia, pp. 4861–4870 (2024)
16. Ma, J., He, Y., Li, F., Han, L., You, C., Wang, B.: Segment anything in medical images. Nat. Commun. **15**(1), 654 (2024)
17. Mei, X., et al.: Radimagenet: an open radiologic deep learning research dataset for effective transfer learning. Radiol. Artifi. Intell. **4**(5), e210315 (2022)
18. Müller, P., Kaissis, G., Zou, C., Rueckert, D.: Joint learning of localized representations from medical images and reports. In: European Conference on Computer Vision, pp. 685–701. Springer (2022). https://doi.org/10.1007/978-3-032-04354-2_17
19. Oquab, M., et al.: Dinov2: Learning robust visual features without supervision. arXiv preprint arXiv:2304.07193 (2023)
20. Radford, A., et al.: Learning transferable visual models from natural language supervision. In: International Conference on Machine Learning, pp. 8748–8763. PMLR (2021)
21. Rajpurkar, Pet al.: Mura: Large dataset for abnormality detection in musculoskeletal radiographs. arXiv preprint arXiv:1712.06957 (2017)
22. Wang, X., Peng, Y., Lu, L., Lu, Z., Bagheri, M., Summers, R.M.: Chestx-ray8: hospital-scale chest x-ray database and benchmarks on weakly-supervised classification and localization of common thorax diseases. In: Proceedings of the IEEE conference on computer vision and pattern recognition, pp. 2097–2106 (2017)
23. Wang, Z., Wu, Z., Agarwal, D., Sun, J.: Medclip: Contrastive learning from unpaired medical images and text. arXiv preprint arXiv:2210.10163 (2022)
24. Wei, X., Vagena, Z., Kurtz, C., Cloppet, F.: Integrating expert knowledge with vision-language model for medical image retrieval. In: 2024 IEEE International Symposium on Biomedical Imaging (ISBI), pp. 1–4. IEEE (2024)
25. Wei, Z., Jin, K., Zhou, X.: Masked contrastive reconstruction for cross-modal medical image-report retrieval. arXiv preprint arXiv:2312.15840 (2023)
26. You, K., et al.: Cxr-clip: toward large scale chest x-ray language-image pre-training. In: International Conference on Medical Image Computing and Computer-Assisted Intervention. pp. 101–111. Springer (2023). https://doi.org/10.1007/978-3-031-43895-0_10
27. Zhang, S et al.: Biomedclip: a multimodal biomedical foundation model pretrained from fifteen million scientific image-text pairs. arXiv preprint arXiv:2303.00915 (2023)
28. Zhang, Y., Jiang, H., Miura, Y., Manning, C.D., Langlotz, C.P.: Contrastive learning of medical visual representations from paired images and text. In: Machine Learning for Healthcare Conference, pp. 2–25. PMLR (2022)
29. Zhou, H.Y., Chen, X., Zhang, Y., Luo, R., Wang, L., Yu, Y.: Generalized radiograph representation learning via cross-supervision between images and free-text radiology reports. Nat. Mach. Intell. **4**(1), 32–40 (2022)

Minimum Data, Maximum Impact: 20 Annotated Samples for Explainable Lung Nodule Classification

Luisa Gallée[1,2,3]($\boxtimes$)(iD), Catharina Silvia Lisson[2](iD), Christoph Gerhard Lisson[2], Daniela Drees[2], Felix Weig[2], Daniel Vogele[2](iD), Meinrad Beer[2,3](iD), and Michael Götz[1,2,3](iD)

[1] Experimental Radiology, Ulm University Medical Center, Ulm, Germany
`luisa.gallee@uni-ulm.de`
[2] Department of Diagnostic and Interventional Radiology, Ulm University Medical Center, Ulm, Germany
[3] XAIRAD - Cooperation for Artificial Intelligence in Experimental Radiology, Ulm, Germany

Abstract. Classification models that provide human-interpretable explanations enhance clinicians' trust and usability in medical image diagnosis. One research focus is the integration and prediction of pathology-related visual attributes used by radiologists alongside the diagnosis, aligning AI decision-making with clinical reasoning. Radiologists use attributes like shape and texture as established diagnostic criteria and mirroring these in AI decision-making both enhances transparency and enables explicit validation of model outputs. However, the adoption of such models is limited by the scarcity of large-scale medical image datasets annotated with these attributes. To address this challenge, we propose synthesizing attribute-annotated data using a generative model. We enhance the Diffusion Model with attribute conditioning and train it using only 20 attribute-labeled lung nodule samples from the LIDC-IDRI dataset. Incorporating its generated images into the training of an explainable model boosts performance, increasing attribute prediction accuracy by 13.4% and target prediction accuracy by 1.8% compared to training with only the small real attribute-annotated dataset. This work highlights the potential of synthetic data to overcome dataset limitations, enhancing the applicability of explainable models in medical image analysis.

Keywords: Explainable AI · Generative AI · Diffusion Model

1 Introduction

When working with a medical dataset containing diagnosis labels, such as malignant or benign tumors, a straightforward approach is to train a classification model to predict tumor malignancy. However, for physicians, merely predicting

© The Author(s), under exclusive license to Springer Nature Switzerland AG 2026
M. Reyes et al. (Eds.): iMIMIC 2025, LNCS 16464, pp. 85–95, 2026.
https://doi.org/10.1007/978-3-032-17611-0_9

a disease is insufficient to establish trust in the model and consequently adopt it in clinical practice. Instead, a more trustworthy approach involves integrating the same diagnostic criteria that radiologists use when making decisions into the model. *Why does the model classify the tumor as malignant? Which criteria does it consider fulfilled?* To facilitate such discussion, the model's training dataset must include annotations for these decision criteria alongside the disease labels. In radiological diagnosis, these criteria correspond to visually identifiable attributes of the pathology, including appearance characteristics such as shape and texture. However, obtaining such annotations requires additional effort, as they are not routinely recorded in standard radiology reports. As a result, large-scale datasets with these annotations are scarce, limiting their applicability for explainable deep learning models.

In this paper, we explore a method for handling such incomplete datasets, using the LIDC-IDRI dataset [25] to classify the malignancy of lung nodules while incorporating their visual characteristics as reasoning attributes. We evaluate a two-step approach: first, augmenting medical images using a generative model conditioned on the attributes; second, adding these synthetic images to the training of explainable image classification models, thereby mitigating the scarcity of human-annotated images, as depicted in Fig. 1.

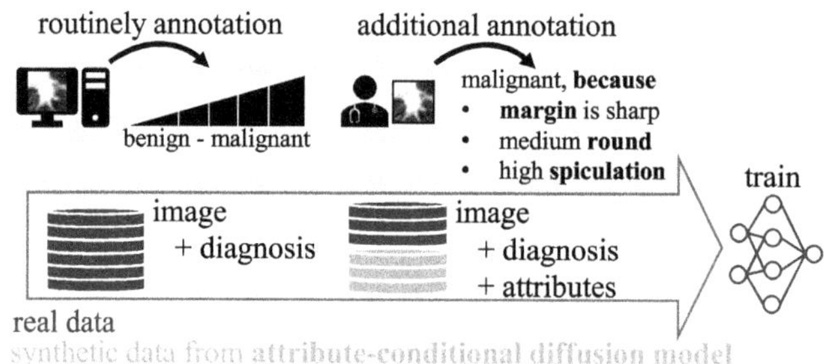

Fig. 1. Large clinical datasets with disease labels are feasible to extract from routine data. However, diagnostic attributes require extra annotation, limiting dataset size. Synthetic augmentation can expand annotated datasets, enabling models to learn attribute-based reasoning in disease prediction.

The generation of medical images has been explored in research for various motivations, including reducing the cost and time required for data acquisition, enabling digital twin technology, handling diverse patient populations, representing multimodal datasets, and ensuring patient data privacy [1–5]. Related work most relevant to our task, focusing on generative AI for lung nodule images, utilizes some nodule characteristics such as size [6] as conditioning criteria to improve the quality of the generated images [7]. Research on using generated images for training classification models, both with attribute conditioning [10,11] and without [8,9], demonstrates performance benefits when these images are used as pre-training data. While some attributes of lung nodules have been incorporated into generative AI models, their use has been limited to improving image quality for disease classification, without providing any insight into the diagnostic reasoning behind the predictions.

Our focus is on investigating how recent advancements in generative models can support the development of explainable AI algorithms. We generate images that satisfy combinations of multiple non-binary attributes and use them to train explainable downstream classification models. The explainability of these models includes both target prediction and attribute prediction.

Methodologically, we employ a Diffusion Model for image synthesis [12]. Compared to Generative Adversarial Networks (GANs) and Variational Autoencoders (VAEs), Diffusion Models have demonstrated superior robustness and performance [14–16]. The conditioning capabilities of Diffusion Models were demonstrated by Rombach et al. [13], who introduced cross-attention mechanisms for conditioning on various modalities. Prior research on multi-label conditioning in Diffusion Models has primarily focused on binary [17] or nominal [18] attributes in general-domain images. The novelties presented in this work are summarized as follows:

- **Attribute-Conditional Generative AI**: We present a Diffusion Model with conditioning on comprehensive diagnostic criteria (attributes) for medical image synthesis.
- **Semi-Conditional Training**: Using labeled and unlabeled data to train the Diffusion Model improves the quality of generated images.
- **Substituting Sparse Datasets**: We leverage the generated annotated data to train SOTA explainable classification models, improving performance on a sparsely annotated subset of the LIDC-IDRI medical benchmark dataset.

The code is publicly available at https://github.com/XRad-Ulm/Attribute-Conditional-Diffusion-Model.

2   Methods

2.1   Attribute-Conditional Diffusion Model

For this work, we employ an attribute-conditional Diffusion Model that processes an image alongside conditioning inputs in the form of multiple metric attributes. The model generates new images that satisfy specific attribute combinations, allowing for the synthesis of an attribute-annotated dataset. The architecture builds upon an existing unconditional Diffusion Model introduced by Ho et al. [12]. We utilize a convolution-based U-Net, structured with two encoder blocks, a central bottleneck, and two decoder blocks. Each block incorporates residual bottlenecks, following the design described by Ma et al. [19,21].

The conditioning mechanism is applied within the network's bottleneck. The attribute vector comprising a single continuous value in the range $[0, 1]$ for each attribute (margin, sphericity, spiculation, etc.) is first projected onto the bottleneck dimensionality via a linear transformation. This transformed attribute latent vector is then integrated with the image data's latent representation through a cross-attention mechanism [20], employing a single-token cross-attention approach without residual connections to reduce parameter count.

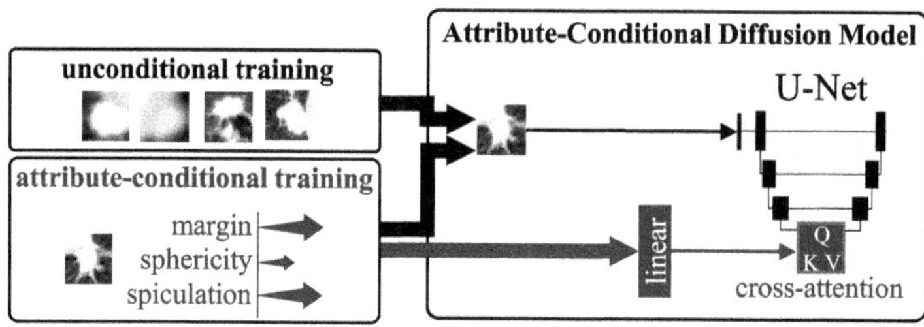

Fig. 2. Generative Model A Diffusion Model for synthesis of attribute-labeled images. Semi-conditional training incorporates both labeled and unlabeled samples by dynamically activating or deactivating cross-attention.

Semi-conditional Training. The cross-attention mechanism provides flexibility for semi-conditional training (see Fig. 2). On one hand, attribute conditioning is achieved by passing the latent vector from cross-attention to the decoder. On the other hand, samples without attribute labels are processed without the attention; their latent representations are directly passed to the decoder. This semi-conditional learning approach is motivated by the limited availability of attribute-labeled samples, which alone may be insufficient for Diffusion Models to generate high-quality images. Unlabeled images, which are relatively easy to obtain in clinical routine, can effectively contribute to this process.

2.2   Attribute-Explainable Classification Model

To evaluate the effect of our synthetic dataset on explainable AI performance, we train two hierarchical classifiers with it, HierViT [22] and the Concept Bottleneck Model [23]. Both architectures first predict visual attributes and then use those attributes to classify the target. HierViT further enhances interpretability by displaying prototypes, representative samples that illustrate each attribute prediction, and generating attribute attention heatmaps. We selected these models because they achieve state-of-the-art accuracy in both attribute and target prediction and offer comprehensive, human interpretable explanations, which have been shown to foster trust among medical professionals [24].

3   Experiments

3.1   Dataset

The Lung Image Database Consortium and Image Database Resource Initiative (CC BY 3.0) [25] provides a densely annotated CT dataset of patients with non-small cell lung cancer. Up to four radiologists segmented lung nodules and rated malignancy on a scale from 1 to 5 [26]. Additionally, the attributes, namely subtlety, internal structure, calcification, sphericity, margin, lobulation, spiculation, and texture, are also annotated using multi-level rating scales. The attributes are well-established diagnostic criteria for assessing the malignancy of lung nodules.

Preprocessing is performed in the same manner as in the comparative experiments described in Gallée et al. [22], which involves excluding nodules identified by fewer than three radiologists or those smaller than 3 mm. Cropouts are generated using the smallest square bounding box and resized to 32×32 pixels for the Diffusion Model, and to 224×224 pixels for the HierViT model using `pylidc` [27]. The dataset consists of 27,379 samples and is evaluated using 5-fold stratified cross-validation by patient, with 10% of the training data reserved for validation.

To reflect our application scenario of sparsely attribute annotated data, only a subset r from the real LIDC IDRI dataset contains additional attribute annotations and is used to train the generative model. In semi-conditional mode the generative model also leverages all remaining images without annotations. For subsequent classification model training, the synthetic attribute-annotated data s is additionally included.

3.2 Implementation Details

Training of Generative Model. Training of the semi-conditional model until loss convergence after 100 epochs on a GeForce RTX 3090 graphics card yielded an average runtime of 40 min. For the training of the classification model synthetic samples were generated by combining real attribute values. Preliminary studies showed that sampling images with random, unrealistic attribute combinations led to poor image quality. The generated samples are also annotated with the respective ground truth malignancy score, as implied by the attribute combination.

Training Classification Model. The model and training setup of HierViT are adopted from the original work [22]. For the Concept Bottleneck Model (CB), we adopt the 'joint' configuration to tightly integrate attribute estimation and target classification. The architecture comprises four fully connected layers with output sizes `[256, 128, 8, 5]`. The third fc-layer generates predictions for the eight visual attributes, and the final layer produces classification scores for the five target classes.

Semi-conditional attribute learning is facilitated by selectively activating or deactivating the attribute loss term, depending on whether the samples contain attribute labels. The real dataset and the synthetic dataset are randomly shuffled for training.

Neither the Attribute-Conditional Diffusion Model nor the classification models were pretrained.

4 Results

4.1 Image Synthesis

User Study. To determine whether experts could distinguish real from synthetic images, we asked participants to view fifty image pairs. Each pair contained one real image and one generated image and participants indicated which image they believed to be real. The generated images were produced by the attention-conditional diffusion models trained on different data volumes. The survey imposed no time limits and did not record response times. Seven domain experts from the University Hospital of Ulm including four senior radiologists with over ten years of experience and three less experienced clinicians participated. Experts identified real images more reliably when generation quality was low but their accuracy declined as quality improved. Less experienced participants showed similarly low discrimination performance regardless of the amount of training data. These results suggest that experienced radiologists can detect model shortcomings when training data are limited yet struggle to distinguish real from synthetic images once image quality improves (Fig. 3).

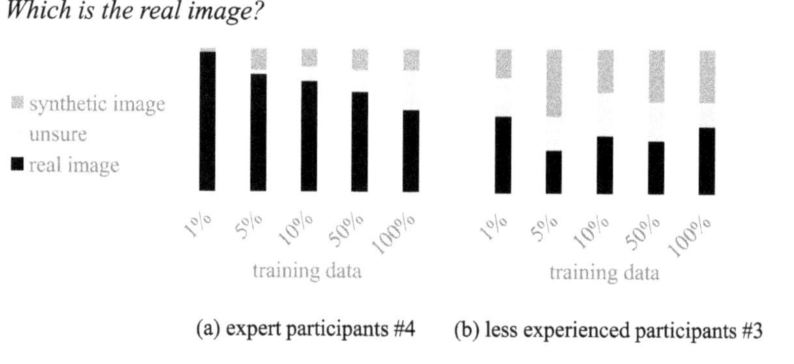

(a) expert participants #4   (b) less experienced participants #3

Fig. 3. Results of the user study assessing the realism of generated images. The plots show the choice made by (a) experts and (b) less experienced participants when asked to identify the real image in a pair. The x-axis denotes the proportion of training data used to train the Attribute-Conditional Diffusion Model.

Quantitative Results. Figure 4 presents examples and quantitative metrics. As attribute labeled training data decreases, SSIM declines while FID and LPIPS increase, reflecting the model's high data dependency. Training with only 20 annotated images yields noisy outputs. Adding semi-conditional training with the remaining images as unconditional data restores SSIM to levels seen with 50–100% conditional data and brings FID and LPIPS into the range of 1–50% conditional data. Training without any attribute information lags behind attribute-conditioned training, underscoring the importance of attribute guidance.

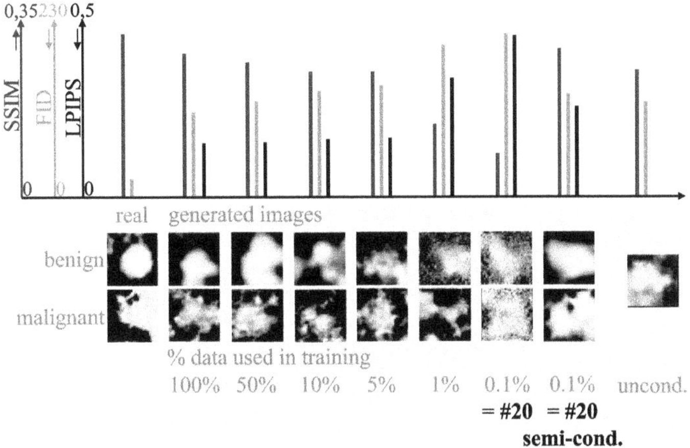

Fig. 4. Qualitative and quantitative assessment of generated data: Structural Similarity Index Measure (SSIM), Fréchet inception distance (FID), Learned Perceptual Image Patch Similarity (LPIPS). The baseline FID is established by splitting the real test set equally into "real" and "fake" distributions.

4.2   Classification Performance

Table 1 reports classification model performance with synthetic images incorporated into the training set. The respective first rows represent the straightforward approach of training the classification models using only the limited attribute-annotated real samples and no synthetic data ($r = 20$, $s = 0$). As expected, both HierViT and CB exhibit low attribute prediction accuracy with high standard deviations. However, HierViT outperforms CB in attribute prediction, likely due to its prototype-based learning approach, which enhances stability under conditions of limited training data [28]. Remarkably, the target accuracy of HierViT is lower compared to the scenario with high attribute availability (last row), despite the use of the same large number of target labels. This suggests that insufficient attribute-annotated data negatively impacts target prediction performance of HierViT. Expanding the training dataset with $s = 2000$ attribute-labeled synthetic samples, as seen in rows two and three, leads to an improvement in HierViT's performance despite the limited availability of real attribute-labeled training data. Semi-conditional training of the Diffusion Model further enhances performance. For CB, both attribute and target prediction performances exhibit high variability and inconsistent results, indicating limited robustness of the model. Further ablation studies on HierViT evaluate the impact of more real and synthetic data ratios on model performance. With $r = 200, s = 2000$ samples, performance (93.2 % mean attribute, 92.8 % target) remains comparable to using $r = 20$ but at a higher annotation cost. Using $r = 20, s = 20'000$ samples lowers performance (91.9 % mean attribute, 90.3 % target), suggesting that excessive synthetic data may introduce noise and reduce classification accuracy.

Table 1. Classification Performance Results are reported in the Within-1-Accuracy metric (%) in mean (black) and standard deviation (gray). A 95% binomial confidence interval is provided for the attribute mean and target. Semi-conditional training of the Diffusion Model is marked with an asterisk *.

attribute data		attributes									target
r	s	sub	is	cal	sph	mar	lob	spic	tex	Ø	malignancy
HierViT [22]											
20	0	85.1	99.8	74.2	79.0	80.3	80.0	85.1	57.3	80.1 [79.6,80.6]	90.9 [90.6,91.2]
		13.3	0.3	32.4	19.3	11.6	15.4	5.4	41.4		3.6
20	2000	94.2	99.4	91.3	96.2	89.7	90.4	88.8	92.5	92.9 [92.6,93.2]	92.0 [91.7,92.3]
		4.1	0.3	5.5	1.7	2.9	3.6	2.7	2.4		1.3
20*	2000	97.0	99.8	90.0	97.2	91.3	91.0	89.2	92.5	93.5 [93.2,93.8]	92.7 [92.4,93.0]
		0.8	0.3	5.4	1.1	1.7	4.1	0.8	1.4		1.6
19713	0	96.3	99.8	95.5	97.4	92.7	94.3	90.8	93.3	95.0 [94.7,95.3]	94.8 [94.5,95.1]
		0.9	0.3	2.1	0.8	1.7	2.9	1.4	1.4		1.4
CB [23]											
20	0	10.2	20.2	24.1	17.1	25.7	25.7	12.0	18.0	19.1 [18.6,19.6]	91.4 [91.1,91.7]
		5.1	7.0	8.9	12.4	9.8	10.5	7.5	7.4		1.4
20	2000	20.4	28.7	26.7	34.0	12.4	37.6	29.2	26.7	27.0 [26.5,27.5]	77.2 [76.7,77.7]
		39.1	34.7	34.2	43.4	25.9	33.9	33.5	38.9		3.8
20*	2000	42.5	69.9	33.3	55.9	41.2	53.6	49.0	43.9	49.5 [48.9,50.1]	78.1 [77.6,78.6]
		39.2	36.0	33.2	31.6	33.1	33.9	35.3	32.9		10.4
19713	0	94.7	99.0	93.9	97.0	94.6	89.0	92.4	92.6	94.2 [93.3,94.7]	91.6 [91.3,91.9]
		1.5	1.5	2.3	2.5	0.9	2.3	3.8	1.0		1.6

5   Discussion and Conclusion

We propose using generative AI to create a richly annotated medical-imaging dataset, and show that adding these synthetic images improves the performance of explainable classification models. Our motivation is twofold: (1) incorporating domain knowledge, such as radiological diagnostic criteria, to enable human-like reasoning and foster trust in the model, and (2) overcoming the scarcity of annotated datasets that include such rich diagnostic metadata.

To this end, we propose an attribute-conditional diffusion model that employs a cross-attention mechanism to guide image generation based on non-binary radiological attributes. Applied to the LIDC-IDRI dataset, our model produces images with high subjective realism, as supported by a user study. While quantitative differences remain compared to real data, even a limited number of labeled real samples (for example, 20 real samples) was sufficient for the synthetic data to improve the performance of a downstream explainable classification model.

In particular, incorporating 2000 synthetic images improved attribute prediction accuracy by 13.4% and target prediction accuracy by 1.8% for the HierViT model, demonstrating the practical value of our method. Although a performance gap persists when compared to a real, fully annotated large-scale dataset (#19713), such attribute-annotated datasets are difficult to obtain due to the substantial annotation effort required.

Our findings highlight the potential of domain-knowledge-driven generative AI to support the development of more accurate and interpretable classification systems in data-scarce medical contexts. Future work could explore extending this approach to other explainability modalities, such as generating paired textual explanations, and integrating them into multi-modal AI pipelines.

Acknowledgments. We thank all participants of the University Hospital Ulm for their contribution to the user study. This research was supported by the German Federal Ministry of Research, Technology and Space (BMFTR) within RACOON COMBINE "NUM 2.0" (FKZ: 01KX2121).

Disclosure of Interests. The authors have no competing interests to declare that are relevant to the content of this article.

References

1. Pezoulas, V.C., Zaridis, D.I., Mylona, E., et al.: Synthetic data generation methods in healthcare: a review on open-source tools and methods. Comput. Struct. Biotechnol. J. (2024). https://doi.org/10.1016/j.csbj.2024.07.005
2. Giuffrè, M., Shung, D.L.: Harnessing the power of synthetic data in healthcare. NPJ Digit. Med. **6**, 186 (2023). https://doi.org/10.1038/s41746-023-00927-3
3. Kazeminia, S., Baur, C., Kuijper, A., et al.: GANs for medical image analysis. Artif. Intell. Med. **109**, 101938 (2020). https://doi.org/10.1016/j.artmed.2020.101938
4. DuMont Schütte, A., Hetzel, J., Gatidis, S., et al.: Overcoming barriers to data sharing with medical image generation: a comprehensive evaluation. NPJ Digit. Med. **4**, 141 (2021). https://doi.org/10.1038/s41746-021-00507-3
5. Zhou, Z., Guo, Y., Tang, R., et al.: Privacy enhancing and generalizable deep learning with synthetic data for mediastinal neoplasm diagnosis. NPJ Digit. Med. **7**, 293 (2024). https://doi.org/10.1038/s41746-024-01290-7
6. Nishio, M., Muramatsu, C., Noguchi, S., et al.: Attribute-guided image generation of three-dimensional computed tomography images of lung nodules using a generative adversarial network. Comput. Biol. Med. **126**, 104032 (2020). https://doi.org/10.1016/j.compbiomed.2020.104032
7. Havaei, M., Mao, X., Wang, Y., Lao, Q.: Conditional generation of medical images via disentangled adversarial inference. Med. Image Anal. **72**, 102106 (2021). https://doi.org/10.1016/j.media.2021.102106
8. Onishi, Y., Teramoto, A., Tsujimoto, M., et al.: Multiplanar analysis for pulmonary nodule classification in CT images using deep convolutional neural network and generative adversarial networks. Int. J. Comput. Assisted Radiol. Surg. **15**, 173–178 (2020). https://doi.org/10.1007/s11548-019-02092-z
9. Onishi, Y., Teramoto, A., Tsujimoto, M., et al.: Investigation of pulmonary nodule classification using multi-scale residual network enhanced with 3DGAN-synthesized volumes. Radiol. Phys. Technol. **13**, 160–169 (2020). https://doi.org/10.1007/s12194-020-00564-5

10. Wang, Q., Zhang, X., Zhang, W., et al.: Realistic lung nodule synthesis with multi-target co-guided adversarial mechanism. Trans. Med. Imaging **40**, 2343–2353 (2021). https://doi.org/10.1109/TMI.2021.3077089
11. Toda, R., Teramoto, A., Tsujimoto, M., et al.: Synthetic CT image generation of shape-controlled lung cancer using semi-conditional InfoGAN and its applicability for type classification. In: Synthetic CT Image Generation of Shape-Controlled Lung Cancer Using Semi-conditional InfoGAN and its Applicability for Type Classification, vol. 16, pp. 241–251 (2021). https://doi.org/10.1007/s11548-021-02308-1
12. Ho, J., Jain, A., Abbeel, P.: Denoising diffusion probabilistic models. In: Advances in Neural Information Processing Systems, vol. 30, pp. 6840–6851 (2020)
13. Rombach, R., Blattmann, A., Lorenz, D., et al.: High-resolution image synthesis with latent Diffusion Models. In: Proceedings of the IEEE/CVF Conference on Computer Vision and Pattern Recognition, pp. 10684–10695 (2022)
14. Goodfellow, I., Pouget-Abadie, J., Mirza, M., et al.: Generative adversarial nets. In: Advances in Neural Information Processing Systems, vol. 27 (2014)
15. Kingma, D.P.: Auto-encoding variational bayes. arXiv preprint arXiv:1312.6114 (2013)
16. Zhan, Z., Chen, D., Mei, J.P., et al.: Conditional image synthesis with diffusion models: a survey. arXiv preprint arXiv:2409.19365 (2024)
17. Lisanti, G., Giambi, N.: Conditioning Diffusion Models via attributes and semantic masks for face generation. Comput. Vis. Image Underst. **244**, 104026 (2024). https://doi.org/10.1016/j.cviu.2024.104026
18. Kong, C., Jeon, D., Kwon, O., Kwak, N.: Leveraging off-the-shelf Diffusion Model for multi-attribute fashion image manipulation. In: Proceedings of the IEEE/CVF Winter Conference on Applications of Computer Vision, pp. 848–857 (2023)
19. Ma, N., Zhang, X., Zheng, H.-T., Sun, J.: Shufflenet v2: practical guidelines for efficient CNN architecture design. In: Proceedings of the European Conference on Computer Vision (ECCV), pp. 116–131 (2018)
20. Vaswani, A.: Attention is all you need. In: Advances in Neural Information Processing Systems (2017)
21. Guocheng, T.: MNISTDiffusion. GitHub repository (2022). https://github.com/bot66/MNISTDiffusion
22. Gallée, L., Lisson, C.S., Beer, M., Götz, M.: Hierarchical vision transformer with prototypes for interpretable medical image classification. arXiv preprint (2025) https://doi.org/10.48550/arXiv.2502.08997
23. Koh, P.W., Nguyen, T., Tang, Y.S., et al.: Concept bottleneck models. In: Proceedings of the International conference on machine learning (ICML), pp. 5338–5348 (2020). https://doi.org/10.5555/3524938.3525433
24. Gallée, L., Lisson, C.S., Lisson, C.G., et al.: Evaluating the explainability of attributes and prototypes for a medical classification model. In: Proceedings of the World Conference on Explainable Artificial Intelligence (xAI), pp. 43–56 (2024). https://doi.org/10.1007/978-3-031-63787-2_3
25. Armato, S.G., III., McLennan, G., Bidaut, L., et al.: Data from LIDC-IDRI. In: The Cancer Imaging Archive (2015). https://doi.org/10.7937/K9/TCIA.2015.LO9QL9SX
26. Armato, S.G., III., McLennan, G., Bidaut, L., et al.: The lung image database consortium (LIDC) and image database resource initiative (IDRI): a completed reference database of lung nodules on CT scans. Med. Phys. **38**, 915–931 (2011). https://doi.org/10.1118/1.3528204

27. Hancock, M.C., Magnan, J.F.: Lung nodule malignancy classification using only radiologist-quantified image features as inputs to statistical learning algorithms: probing the Lung Image Database Consortium dataset with two statistical learning methods. J. Med. Imaging **3**, 044504 (2016). https://doi.org/10.1117/1.JMI.3.4.044504
28. Snell, J., Swersky, K., Zemel, R.: Prototypical networks for few-shot learning. In: Advances in Neural Information Processing Systems, vol. 30 (2017)

Evaluating the Explainability of Vision Transformers in Medical Imaging

Leili Barekatain$^{(\boxtimes)}$ and Ben Glocker

Department of Computing, Imperial College London, London, UK
`lb124@imperial.ac.uk`

Abstract. Understanding model decisions is crucial in medical imaging, where interpretability directly impacts clinical trust and adoption. Vision Transformers (ViTs) have demonstrated state-of-the-art performance in diagnostic imaging; however, their complex attention mechanisms pose challenges to explainability. This study evaluates the explainability of different Vision Transformer architectures and pre-training strategies—ViT, DeiT, DINO, and Swin Transformer—using Gradient Attention Rollout and Grad-CAM. We conduct both quantitative and qualitative analyses on two medical imaging tasks: peripheral blood cell classification and breast ultrasound image classification. Our findings indicate that DINO combined with Grad-CAM offers the most faithful and localized explanations across datasets. Grad-CAM consistently produces class-discriminative and spatially precise heatmaps, while Gradient Attention Rollout yields more scattered activations. Even in misclassification cases, DINO with Grad-CAM highlights clinically relevant morphological features that appear to have misled the model. By improving model transparency, this research supports the reliable and explainable integration of ViTs into critical medical diagnostic workflows. The code is available at https://github.com/leilibrk/Explainability-of-ViTs.

Keywords: Vision Transformers · Explainability · Medical Imaging · Grad-CAM · Gradient Attention Rollout

1 Introduction

The integration of artificial intelligence into medical image analysis has led to significant advancements, particularly in biomedical image classification. Precise classification of medical images helps in early disease detection and reduces the risk of misdiagnosis. Although deep learning models excel in general image classification, biomedical images present challenges due to complex structures, varying noise levels, and intra- and inter-class variability. Convolutional Neural Networks (CNNs) have played a pivotal role in this domain by learning hierarchical features, yet struggle to capture long-range dependencies within images.

To overcome these limitations, Vision Transformers (ViTs) have emerged as a promising alternative, leveraging self-attention mechanisms to model global

© The Author(s), under exclusive license to Springer Nature Switzerland AG 2026
M. Reyes et al. (Eds.): iMIMIC 2025, LNCS 16464, pp. 96–105, 2026.
https://doi.org/10.1007/978-3-032-17611-0_10

relationships across image patches. Unlike CNNs, ViTs process entire images holistically, leading to improved performance in complex classification tasks. However, a key challenge remains: the interpretability of these models.

Explainability is critical in medical applications, where clinicians must validate model predictions to ensure they are based on meaningful biological features rather than spurious correlations. Relying solely on performance metrics such as accuracy is insufficient; we must also apply explainability techniques to understand and trust model decisions. Current interpretability techniques, such as attention-based and feature attribution methods, provide insights into model decision-making but vary in effectiveness across different architectures.

This study aims to bridge this gap by evaluating different Vision Transformer architectures and pre-training strategies, namely ViT, DeiT, DINO, and Swin Transformer on peripheral blood cell classification and breast ultrasound image classification tasks. We assess their explainability using Gradient Attention Rollout and Grad-CAM, analyzing their ability to provide reliable, interpretable predictions. We investigate (1) how effectively these models classify images and (2) how well different explainability methods reveal their decision-making process. Our findings contribute to the growing field of AI-driven medical diagnostics by shedding light on the balance between model accuracy and interpretability.

1.1   Related Work

Recent works have provided valuable insights into making Vision Transformers interpretable in the context of medical imaging. Komorowski et al. [12] introduced metrics—faithfulness, sensitivity, and complexity—to benchmark model explainability. They conducted experiments on chest X-ray classification, comparing methods like LIME [16], Attention Rollout [1], and TransLRP [5]. However, their analysis was limited to fine-tuning a base-sized ViT model and did not explore how different ViT architectures may affect explainability.

Several studies have also developed explainable ViTs for specific medical tasks. xViTCOS [14] is one such model, designed for COVID-19 screening using chest X-rays. It employs Gradient Attention Rollout [8] to assess model explainability. The results show that the model not only outperforms recent benchmarks but also attends to meaningful image regions, as confirmed by radiologists. However, the study assesses explainability qualitatively through visual alignment with clinical expectations and does not provide quantitative metrics to objectively compare the attribution quality of the generated saliency maps. A related study [9] also focused on COVID-19 and pneumonia prediction from chest X-rays using different explainable deep learning models, including ViTs. The system integrates Grad-CAM [17] to visualize critical regions influencing predictions and reports high diagnostic accuracy across multiple lung disease categories. However, it also lacked a quantitative evaluation of explainability.

2   Methods

The **Vision Transformer (ViT)** [7] divides an image into fixed-size non-overlapping patches, embeds them linearly, and processes them using a transformer encoder to capture global context. Each encoder block uses self-attention, which allows every patch to attend to all others. This enables the model to learn both local and global dependencies. ViTs have shown strong performance in medical imaging tasks such as disease classification and lesion detection, often outperforming CNNs. However, standard ViTs often require large labeled datasets, which can limit their use in medical applications.

To reduce the reliance of ViTs on large-scale datasets, the **Data-efficient Image Transformer (DeiT)** was proposed [19]. DeiT uses a student-teacher distillation approach, where a CNN teacher guides training via a distillation token that interacts with class and patch tokens. This helps the model learn from the teacher's predictions and achieve strong performance on smaller datasets. Sevinc et al. [18] utilized DeiT architecture for classifying brain MRI images using a small dataset of 3,264. The model achieved a notable accuracy of 93.69%.

The **DINO (Self-Distillation with No Labels)** framework [4] is a self-supervised approach that trains ViT without labels using a teacher–student model, where both networks share the same architecture and process different augmented views. The teacher is updated using an exponential moving average of the student, allowing the model to learn meaningful visual representations without explicit supervision. Cisternino et al. [6] used DINO to classify histopathological images across 23 tissue types. The model extracted meaningful features without manual labels and outperformed other self-supervised methods.

Swin Transformer [13] improves ViTs by introducing hierarchical feature representation and shifted window attention, reducing computational complexity from quadratic to linear. Self-attention is applied within local windows, and shifting these windows across layers enables global context modeling. Jin et al. [10] proposed the Multitask Swin Transformer (MTST) for pulmonary nodule analysis in CT images. This multitask approach, which predicts malignancy and characterization scores concurrently, provides clinicians information similar to a radiologist's assessment and improves the model's practical diagnostic utility.

2.1   Explainability Techniques

As Vision Transformers rapidly advance medical imaging, understanding their decision-making has become crucial. Here, we discuss two common methods for interpreting ViT decisions.

Attention-Based Methods. Attention-based methods aim to explain Vision Transformers by analyzing how attention is distributed across image tokens. Since attention mechanisms allow the model to assign different importance levels to different regions of the input, they offer a natural way to explore interpretability [20]. One widely used technique is Attention Rollout [1], which tracks how information flows through the transformer layers by recursively multiplying averaged attention matrices:

$$rollout = \hat{A}^{(1)} \cdot \hat{A}^{(2)} \cdot ... \cdot \hat{A}^{(B)} \tag{1}$$

However, Attention Rollout produces the same explanation regardless of the predicted class, which limits its ability to highlight class-specific features [11]. To address this, **Gradient Attention Rollout** [8] weights attention layers using gradients to highlight regions of the image that contributed to the model's decision. To ensure class-specificity, attention values are multiplied by corresponding gradients. The final attention map is obtained by averaging across attention heads, preserving only most relevant and positively contributing attention paths for the target class. In this way, Gradient Attention Rollout produces class-specific attention maps that better reflect the model's decision-making process.

Feature Attribution Methods. Unlike attention-based techniques that analyze the internal model mechanisms, feature attribution methods explain model predictions by identifying which input features most influenced the output. **Grad-CAM** [17] is a gradient-based feature attribution method that computes the gradient of the target class score with respect to the activations of a selected transformer layer, averaging these gradients to get importance weights for each feature map, and combining them to generate a heatmap. ReLU activation keeps only positively contributing features, and the result is upsampled to the input resolution. The final output is a class-discriminative heatmap that highlights regions most responsible for the model's decision. In medical imaging, such visual explanations are essential to ensure models rely on clinically meaningful features.

3   Experiments and Results

3.1   Datasets and Implementation Details

Peripheral Blood Cell (PBC) Dataset [2] is a public dataset that contains 17,092 high-resolution (360×363) JPEG images of eight categories of peripheral blood cells, including Basophil, Eosinophil, Erythroblast, Immature Granulocyte, Lymphocyte, Monocyte, Neutrophil, and Platelet.

Breast Ultrasound Images Dataset [3] is a public dataset containing 780 PNG images (500×500). The data were collected in 2018 from 600 female patients, aged between 25 and 75 years. The images are categorized into three classes: normal, benign, and malignant.

Implementation Details All images were resized to 224×224, center-cropped, and normalized, with no additional preprocessing or augmentation.

We fine-tuned four pre-trained Vision Transformer models from the `PyTorch Transformers` library: ViT (google/vit-base-patch16-224), DeiT (facebook/deit-tiny-patch16-224), DINO-ViT (facebook/dino-vits16), and Swin (microsoft/swin-tiny-patch4-window7-224). Training was performed with a batch size of 32. Given that these models were extensively pre-trained on large-scale datasets, and considering the relatively modest size of the PBC dataset, one epoch was sufficient to fine-tune the models effectively without risking overfitting. However, for the breast ultrasound image classification dataset, due to its smaller size (780 images), we experimented with different settings and found that training the models for 10 epochs was sufficient to achieve stable performance without overfitting. Throughout the training and evaluation phases, we monitored performance metrics, including accuracy and F1-score. After training, Gradient Attention Rollout and Grad-CAM were applied to analyze the explainability of model predictions.

3.2   Performance Results

The performance of the four vision transformer architectures, including ViT, DeiT, DINO-ViT, and Swin was evaluated using standard classification metrics for two datasets. Table 1 and 2 present the results.

Table 1. Results on PBC

Model	Accuracy (%)	F1-score (%)
ViT	98.68	98.73
DeiT	98.05	97.92
DINO	96.97	97.16
Swin	98.58	98.59

Table 2. Results on Breast Ultrasound

Model	Accuracy (%)	F1-score (%)
ViT	87.18	85.66
DeiT	79.49	75.48
DINO	80.77	77.23
Swin	89.74	88.44

3.3   Explainability Results—Quantitative

Insertion and Deletion are complementary metrics for evaluating the faithfulness of an explanation method. In the Insertion metric, the most important pixels from the generated heatmaps are gradually added to a blank or blurred image, and the probabilities of the target class are tracked as pixels are inserted. A steeper increase in confidence indicates that the explanation has accurately identified the most relevant regions. Conversely, in the Deletion metric, the most important pixels are progressively masked from the original image, and the corresponding drop in target class probabilities is recorded. A more rapid decline suggests higher explanation fidelity [11,15].

We evaluated the explainability of Gradient Attention Rollout and Grad-CAM across four architectures using these metrics. For the PBC dataset, we randomly sampled 520 validation images (65 per class). For the Breast Ultrasound dataset, we used all 117 validation images. Both methods were applied uniformly across models and Insertion and Deletion AUC were used to measure how well explanations matched the models' decisions. Results in Fig. 1 show that Grad-CAM consistently outperforms Gradient Attention Rollout across all architectures, showing higher AUC in insertion (faster confidence recovery) and lower AUC in deletion (steeper confidence drop), indicating better localization of critical visual features.

The Insertion and Deletion AUCs are presented in Table 3 and Table 4. Based on the results, DINO gives the best AUC scores for both datasets, suggesting that it is easier to explain this model's decisions using Grad-CAM.

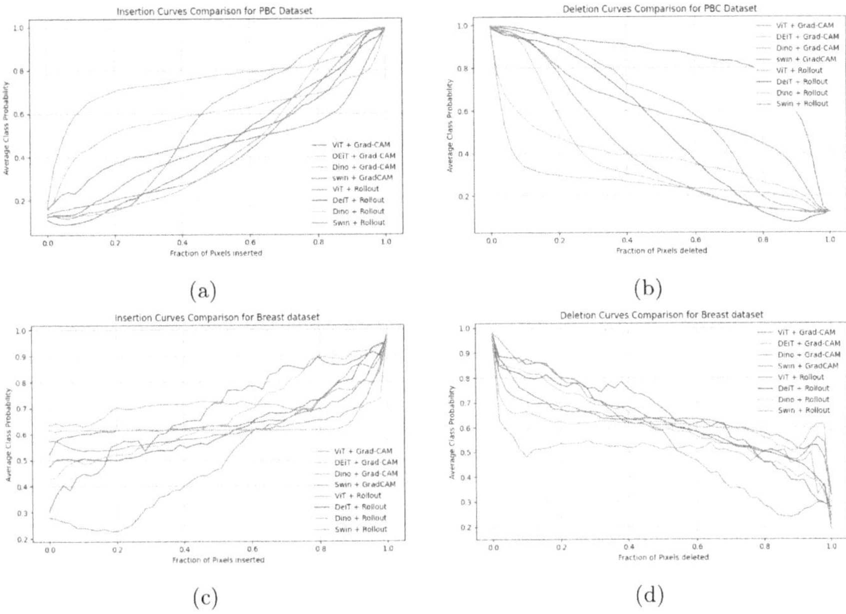

Fig. 1. Visualization of (a, c) inserting and (b, d) deleting the most relevant pixels—identified by Grad-CAM and Gradient Attention Rollout—on the predicted class probability across four Vision Transformer models for the PBC dataset (a, b) and Breast Ultrasound dataset (c, d).

Table 3. The Deletion and Insertion AUC across models using Grad-CAM and Rollout explanations for PBC dataset.

	Grad-CAM				Rollout			
	ViT	DeiT	DINO	Swin	ViT	DeiT	DINO	Swin
Deletion($\downarrow$)	0.60	0.38	**0.27**	0.82	0.42	0.52	0.36	0.60
Insertion($\uparrow$)	0.44	0.60	**0.75**	0.52	0.44	0.45	0.45	0.56

Table 4. The Deletion and Insertion AUC across models using Grad-CAM and Rollout explanations for Breast Ultrasound dataset.

	Grad-CAM				Rollout			
	ViT	DeiT	DINO	Swin	ViT	DeiT	DINO	Swin
Deletion($\downarrow$)	0.63	0.61	**0.51**	0.65	0.61	0.63	0.60	0.56
Insertion($\uparrow$)	0.65	0.67	**0.72**	0.69	0.61	0.62	0.62	0.50

3.4 Explainability Results Qualitative

Figures 2 and 3 show the qualitative results of Gradient Attention Rollout and Grad-CAM across four transformer models on the PBC and breast ultrasound datasets. In general, Grad-CAM produces more focused and interpretable heatmaps, with stronger activation over the relevant regions. However, Gradient Attention Rollout produces scattered and inconsistent attention maps that often highlight irrelevant background regions, making its explanations more difficult to interpret. Among the models, DINO-ViT combined with Grad-CAM consistently provides the most coherent and clinically meaningful attributions.

For example, in Fig. 2 *right* in the Basophil case (first row), DINO with Grad-CAM highlights the entire cell body accurately, while other models show more dispersed or weaker activations. Similarly, in the Immature Granulocyte case (second row), DINO with Grad-CAM shows a precise heatmap centered over the nucleus and cytoplasmic granules.

In Fig. 3 *right*, it is evident that DINO with Grad-CAM again demonstrates superior explainability. In the benign case (first row), it produces a focused activation map centered on the lesion boundary, while other models such as DeiT and ViT display scattered attention across the entire image, including irrelevant tissue regions. Swin shows stronger activation but extends beyond the lesion area. In the malignant case (second row), DINO localizes the irregularly shaped mass, focusing precisely on the contours of the tumor.

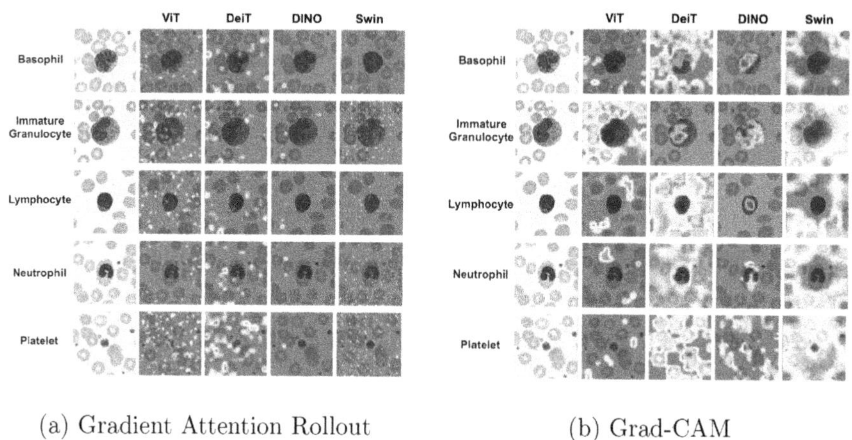

(a) Gradient Attention Rollout   (b) Grad-CAM

Fig. 2. Comparison of Gradient Attention Rollout and Grad-CAM heatmaps for five blood cell classes from the PBC dataset across four models. First column shows the input images, followed by heatmaps of the predicted class.

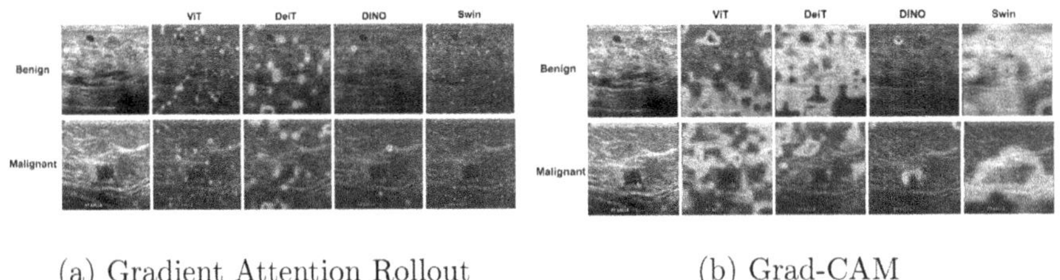

(a) Gradient Attention Rollout   (b) Grad-CAM

Fig. 3. Comparison of Gradient Attention Rollout and Grad-CAM heatmaps for benign and malignant breast ultrasound images across four models. First column shows the input images, followed by heatmaps of the predicted class.

Qualitative Error Analysis. When a model is explainable, its visual attributions—such as Grad-CAM heatmaps—can provide insights into the reasoning behind incorrect predictions. Figure 4 shows two such cases by the DINO-ViT model. In the PBC example (top), the model predicted the class as *Immature Granulocyte* with 67.43% confidence, while the ground truth label is *Monocyte*. The Grad-CAM heatmap for the predicted class shows strong activation around the nucleus, suggesting the model misinterpreted the shape and granularity as irregular features typical of Immature Granulocyte cells, which led to the mis-classification. Meanwhile, the heatmap for the true class displays weaker and more diffused attention, suggesting that the model has not learned a sufficiently strong representation for Monocyte. The activation patterns observed across other classes further demonstrate how subtle visual similarities can lead to confusion in fine-grained classification tasks.

In the breast ultrasound example (bottom), the model misclassified a *Benign* lesion as *Malignant* with 79.17% confidence. The heatmap for the predicted class highlights the central lesion area with strong intensity, indicating that the model focused on structural features associated with malignancy, such as asymmetry or irregular margins. Meanwhile, the heatmap for the true class appears weaker and

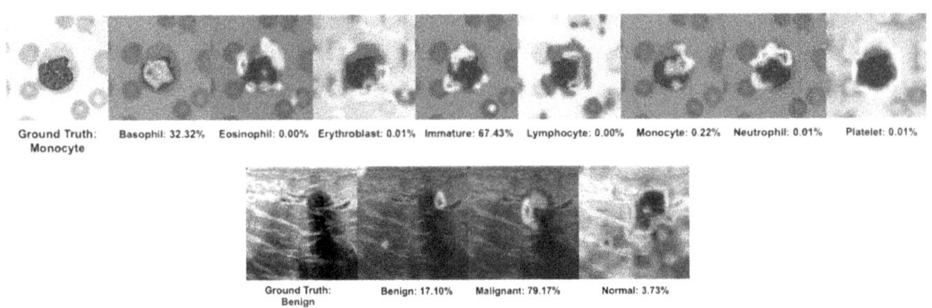

Fig. 4. Grad-CAM visualizations of misclassified images by the DINO-ViT model. **Top:** PBC dataset—ground truth is Monocyte, but predicted as Immature Granulocyte with 67.43% confidence. **Bottom:** Breast ultrasound—ground truth is Benign, but predicted as Malignant with 79.17% confidence.

less defined, suggesting the model failed to learn sufficiently discriminative features for benign cases. These examples demonstrate that explainability methods like Grad-CAM can reveal the underlying reasons behind model misclassifications, offering valuable insights for diagnosing and improving model behavior.

4   Conclusion and Future Works

This study aimed to provide a systematic framework for identifying the most explainable Vision Transformer architecture and explanation method for medical imaging tasks. We evaluated four architectures—ViT, DeiT, DINO-ViT, and Swin Transformer—using two explainability techniques: Grad-CAM and Gradient Attention Rollout. Our experiments on peripheral blood cell (PBC) and breast ultrasound datasets revealed that while all models achieved high classification accuracy, their interpretability varied notably. Grad-CAM consistently provided more localized and class-discriminative explanations compared to Gradient Attention Rollout. When using Grad-CAM as the benchmark for explainability, DINO emerged as the most interpretable setup. In contrast, ViT, DeiT, and Swin Transformer showed scattered and less focused attention, which may limit their clinical applicability. Even in misclassification cases, DINO's attention maps still highlighted relevant morphological features, helping to uncover potential reasons behind the model's decision-making. Interestingly, ViT and Swin outperformed DINO-ViT in terms of accuracy and F1-score across both datasets (see Tables 1 and 2). However, DINO consistently provided more explainable results. This suggests that model selection in critical domains like medical diagnostics should not rely solely on performance metrics, but also consider the quality of explanations.

Future work could focus on developing more accurate and clinically meaningful explainability methods specifically designed for ViTs. Designing hybrid techniques that combine spatial precision with deeper semantic understanding could enhance interpretability. Additionally, incorporating domain-specific priors or medical constraints into the explanation process may improve faithfulness.

Disclosure of Interests. The authors have no competing interests to declare.

References

1. Abnar, S., Zuidema, W.: Quantifying attention flow in transformers. In: Proceedings of the 58th Annual Meeting of the Association for Computational Linguistics (ACL), pp. 4190–4197. Association for Computational Linguistics (2020)
2. Acevedo, A., Merino, A., Alférez, S., Molina, Á., Boldú, L., Rodellar, J.: A dataset of microscopic peripheral blood cell images for development of automatic recognition systems. Data Brief **30**, 105474 (2020)
3. Al-Dhabyani, W., Gomaa, M., Khaled, H., Fahmy, A.: Dataset of breast ultrasound images. Data Brief **28**, 104863 (2020)
4. Caron, M., et al.: Emerging properties in self-supervised vision transformers. arXiv preprint arXiv:2104.14294 (2021)

5. Chefer, H., Gur, S., Wolf, L.: Transformer interpretability beyond attention visualization. In: Proceedings of the IEEE/CVF Conference on Computer Vision and Pattern Recognition, pp. 782–791 (2021)
6. Cisternino, F., Ometto, S., Chatterjee, S., Giacopuzzi, E., Levine, A.P., Glastonbury, C.A.: Self-supervised learning for characterising histomorphological diversity and spatial RNA expression prediction across 23 human tissue types. Nat. Commun. **15**(1), 5906 (2024)
7. Dosovitskiy, A., et al.: An image is worth 16×16 words: transformers for image recognition at scale. arXiv preprint arXiv:2010.11929 (2020)
8. Gildenblat, J.: ViT-Explain: Explainability for Vision Transformers (2021). https://github.com/jacobgil/vit-explain. Accessed 02 June 2025
9. Hroub, N.A., Alsannaa, A.N., Alowaifeer, M., Alfarraj, M., Okafor, E.: Explainable deep learning diagnostic system for prediction of lung disease from medical images. Comput. Biol. Med. **170**, 108012 (2024)
10. Jin, H., Yu, C., Zhang, J., Zheng, R., Fu, Y., Zhao, Y.: Multitask swin transformer for classification and characterization of pulmonary nodules in CT images. Quant. Imaging Med. Surg. **15**(3), 1845 (2025)
11. Kashefi, R., Barekatain, L., Sabokrou, M., Aghaeipoor, F.: Explainability of vision transformers: a comprehensive review and new perspectives. arXiv preprint arXiv:2311.06786 (2023)
12. Komorowski, P., Baniecki, H., Biecek, P.: Towards evaluating explanations of vision transformers for medical imaging. In: Proceedings of the IEEE/CVF Conference on Computer Vision and Pattern Recognition, pp. 3726–3732 (2023)
13. Liu, Z., et al.: Swin transformer: hierarchical vision transformer using shifted windows. In: Proceedings of the IEEE/CVF International Conference on Computer Vision (2021)
14. Mondal, A.K., Bhattacharjee, A., Singla, P., Prathosh, A.: xvitcos: explainable vision transformer based covid-19 screening using radiography. IEEE J. Transl. Eng. Health Med. **10**, 1–10 (2021)
15. Petsiuk, V., Das, A., Saenko, K.: Rise: randomized input sampling for explanation of black-box models. arXiv preprint arXiv:1806.07421 (2018)
16. Ribeiro, M.T., Singh, S., Guestrin, C.: "Why should I trust you?" explaining the predictions of any classifier. In: Proceedings of the 22nd ACM SIGKDD International Conference on Knowledge Discovery and Data Mining, pp. 1135–1144 (2016)
17. Selvaraju, R.R., Cogswell, M., Das, A., Vedantam, R., Parikh, D., Batra, D.: Gradcam: visual explanations from deep networks via gradient-based localization. In: Proceedings of the IEEE International Conference on Computer Vision (ICCV) (2017)
18. Sevinc, A., Ucan, M., Kaya, B.: A distillation approach to transformer-based medical image classification with limited data. Diagnostics **15**(7), 929 (2025)
19. Touvron, H., Cord, M., Sablayrolles, A., Synnaeve, G., Jégou, H.: Training data-efficient image transformers & distillation through attention. arXiv preprint arXiv:2012.12877 (2021)
20. Vaswani, A., et al.: Attention is all you need. In: Advances in Neural Information Processing Systems, vol. 30 (2017)

A Hybrid Fully Convolutional CNN-Transformer Model for Inherently Interpretable Disease Detection from Retinal Fundus Images

Kerol Djoumessi[1]([✉]) [iD], Samuel Ofosu Mensah[1] [iD], and Philipp Berens[1,2]([✉]) [iD]

[1] Hertie Institute for AI in Brain Health, University of Tübingen, Tübingen, Germany
kerol.djoumessi-donteu@uni-tuebingen.de

[2] Tübingen AI Center, University of Tübingen, Tübingen, Germany
philipp.berens@uni-tuebingen.de

Abstract. In many medical imaging tasks, convolutional neural networks (CNNs) efficiently extract local features hierarchically. More recently, vision transformers (ViTs) have gained popularity, using self-attention mechanisms to capture global dependencies, but lacking the inherent spatial localization of convolutions. Therefore, hybrid models combining CNNs and ViTs have been developed to combine the strengths of both architectures. However, such hybrid models are difficult to interpret, which hinders their application in medical imaging. In this work, we introduce an interpretable-by-design hybrid fully convolutional CNN-Transformer architecture for retinal disease detection. Unlike widely used post-hoc saliency methods for ViTs, our approach generates faithful and localized evidence maps that directly reflect the model's decision process. We evaluated our method on two medical tasks focused on disease detection using color fundus images. Our model achieves state-of-the-art predictive performance compared to black-box and interpretable models and provides class-specific sparse evidence maps in a single forward pass.

Keywords: Self-explainability · interpretable-by-design · Hybrid CNN-Transformer · Dual-Resolution Self-Attention · Retinal fundus images

1 Introduction

Convolutional neural networks (CNNs) are at the heart of many successful applications in medical image analysis [9], but more recently, vision transformers (ViTs) have emerged as a competitive alternative [11], demonstrating strong performance in medical imaging tasks [3,26]. Although CNNs are highly effective at capturing complex local patterns in images, the size of their receptive field is smaller than some disease-related lesions [17]. In contrast, vision transformers

© The Author(s), under exclusive license to Springer Nature Switzerland AG 2026

M. Reyes et al. (Eds.): iMIMIC 2025, LNCS 16464, pp. 106–116, 2026.
https://doi.org/10.1007/978-3-032-17611-0_11

leverage self-attention (SA) [27] to capture long-range dependencies, providing a more global understanding of the image. Despite these advantages, ViTs require substantial computational resources, often demanding large-scale datasets for effective training [20,26], while also facing challenges in interpretability [16].

To address the weaknesses of both approaches, a promising alternative are hybrid CNN-Transformer architectures. Several studies have used such architectures [15,18,20,26], improving performance for tasks that require combining local features with global relationships for classification. Yet, the interpretability of such hybrid approaches has remained a challenge [18,20,26], as they require either techniques tailored to transformer architectures or the development of novel visualization methods [18]. To this end, either CNN-based methods have been adapted to ViTs [4,24] or ViT-specific techniques have been proposed [1,7,8]. The most commonly used ViT-specific approach has been to visualize attention maps across layers, as these capture interactions between input regions. However, attention is not class-specific and merely illustrates relationships between input patches rather than their direct contribution to the model prediction [5,16,25]. Alternatively, post-hoc CNN-based methods like LRP [4] and GradCAM [24] have been successfully adapted to ViT by integrating gradients within the self-attention layers, offering class-wise explanations [8]. Yet, these are model-specific and struggle with hierarchical architectures like the Swin Transformer [21].

Here, we propose a novel, inherently interpretable-by-design hybrid CNN-Transformer architecture for retinal fundus image classification, combining the feature extraction strengths of CNNs with the ability of ViTs to capture long-range dependencies from dual-resolution features. Dual-resolution self-attention (DRSA) allows the model to capture both fine-grained details and global context by attending to representations at two distinct spatial resolutions. Our design integrates recent advancements such as convolutional ViTs [29], dual-resolution self-attention [15], and sparse explanations [10,17]. We evaluated our model using two backbone CNNs—ResNet and BagNet—on two clinically relevant tasks: Diabetic Retinopathy (DR) detection and Age-Related Macular Degeneration (AMD) severity classification, using publicly available color fundus image datasets. Our hybrid model provides self-interpretability without sacrificing classification performance, challenging the myth of the accuracy-interpretability tradeoff [23]. It maintained predictive performance compared to both interpretable and non-interpretable state-of-the-art models while offering faithful explanations that accurately localize disease-related lesions—even under distribution shift—outperforming traditional post-hoc methods.

2   Developing a Self-explainable Hybrid CNN-ViT Model

2.1   Hybrid CNN-ViT Architecture

In our hybrid architecture (Fig. 1), CNN and ViT modules are used sequentially, with the output of the CNN module serving directly as the input to the trans-

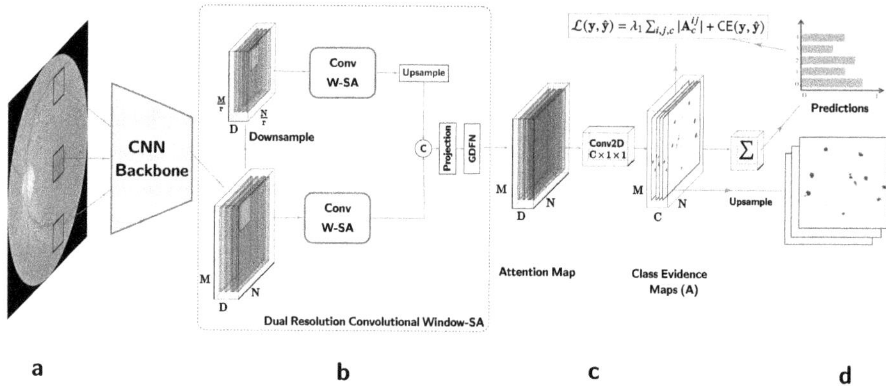

Fig. 1. Interpretable-by-design hybrid CNN-Transformer model. (a) Input image. The black patches illustrate the small receptive field of the CNN backbone (BagNet). (b) Two window-based SA modules are applied separately to the high and downsampled low-resolution feature maps, followed by feature fusion. (c) The high-dimensional attention map is transformed into the class evidence map **A** by applying a 1×1 convolutional classifier with C kernels, where C is the number of classes. (d) Spatial averaging of the class evidence maps yields predictions, whereas upsampling **A** provides explanations.

former module. Specifically, the CNN module acted as a feature extractor, capturing local patterns. The ViT module modeled long-range dependencies between the extracted features, enhancing the model's ability to understand broader contexts. Given an input image $\mathbf{X} \in \mathbb{R}^{H \times W \times C}$—where H, and W, denote height and width, and C is the number of channels—the CNN backbone f extracts a spatial feature representation $\mathbf{Z} = f_\theta(\mathbf{X}) \in \mathbb{R}^{M \times N \times D}$, where θ denotes the model parameter, $M \times N$ represents the spatial size, and D is the feature dimension. We used either a ResNet-50 (with a receptive field of 427×427) or a BagNet-33 (33×33) as the backbone network. Unlike the ResNet, the BagNet aggregates only local features in a bag-of-words manner [6]. The transformer module (Fig. 1b) uses a dual-convolutional window self-attention (Conv-wSA) mechanism that operates on both high- and low-resolution versions of the original feature maps to produce an attention map $\mathbf{W} = g_\phi(\mathbf{Z}_h, \mathbf{Z}_l) \in \mathbb{R}^{M \times N \times D}$. Here, $\mathbf{Z}_h = \mathbf{Z}$ denotes the high-resolution feature map, while $\mathbf{Z}_l = d(\mathbf{Z}, r) \in \mathbb{R}^{\frac{M}{r} \times \frac{N}{r} \times D}$ is the low-resolution counterpart, obtained via the downsampling function $d(\cdot)$ with reduction factor r. The function g_ϕ, parametrized by ϕ, jointly encompasses the parameterized projection and the Gated-Dconv Feed-Forward Network (GDFN) [31]. The attention map produced by the transformer module maintains the spatial resolution of the input feature map. The classification module (Fig. 1c) comprises a convolutional layer with C kernels of size 1×1 and unit stride, producing an evidence map $\mathbf{A} = h_\psi(\mathbf{W}) \in \mathbb{R}^{M \times N \times C}$, where C represents the number of classes and ψ denotes the parameter of the classifier h. The final prediction $\hat{\mathbf{y}} \in \mathbb{R}^{1 \times C}$ is obtained by applying spatial average pooling to $\mathbf{A}$, followed by a softmax operation: $\hat{\mathbf{y}} = \mathrm{Softmax}(\mathrm{AvgPool}(\mathbf{A}))$. This yields a C-dimensional probability distribution representing the likelihood of each class.

2.2 Learning Long-Range Dependencies with Convolutional DRSA

To learn long-range dependencies between the convolutional features, we used a transformer module with dual-resolution self-attention (DRSA) [15], for which a convolutional layer had replaced the linear fully connected layer (FCL) [29] as follows: $\mathrm{SA}_h = \mathrm{Softmax}\left(\frac{\mathbf{Q}_h \mathbf{K}_h^\top}{\alpha}\right)\mathbf{V}_h$, $\mathrm{SA}_l = \mathrm{Softmax}\left(\frac{\mathbf{Q}_l \mathbf{K}_l^\top}{\alpha}\right)\mathbf{V}_l$ where α is the scaling factor, and $\mathbf{Q}_h, \mathbf{K}_h, \mathbf{V}_h$ and $\mathbf{Q}_l, \mathbf{K}_l, \mathbf{V}_l$ are the queries, keys, and value embeddings generated for $\mathbf{Z}_h$ and $\mathbf{Z}_l$ using convolutional operations. The final self-attention representation is computed as: $\mathrm{SA}_{final} = \mathrm{GDFN}_\delta\left(\mathrm{Proj}_\beta(\mathrm{SA}_h + \mathrm{Up}(\mathrm{SA}_l))\right)$, with $\mathbf{W} = \mathrm{SA}_{final}$, where $\mathrm{Up}(\mathrm{SA}_l)$ denotes the upsampled version of SA_l. This upsampled map is aggregated with SA_h and passed through a convolutional projection parametrized by β. The resulting representation is subsequently refined using a GDFN parametrized by δ, which enhances spatial structures while suppressing irrelevant features. This refinement ensures that only salient information contributes to the final predictions, thereby improving the generalization performance of the model.

2.3 Enhancing Interpretability with a Sparse Convolutional Classifier

In standard ViT and hybrid CNN-Transformer models, the classification head includes a FCL, which discards spatial information, limiting interpretability. Our architecture addressed this by preserving spatial information using convolutional operations in the self-attention module, generating attention maps that capture long-range dependencies between regions in the same window. To enhance interpretability, we replaced the FCL with a convolutional classifier, referred to as the *class evidence layer*. This layer leverages spatial information to produce class-wise evidence maps (Fig. 1), where each pixel reflects the local contribution of input regions to the final prediction. Following classification, the evidence maps are upsampled and overlaid on the input image for visualization purposes (Fig. 1d). Furthermore, the inclusion of an explicit class evidence layer enables the application of an ℓ_1 sparsity constraint on the class evidence maps $\mathbf{A}_c$, thereby enhancing interpretability [10,17]. This leads to the following loss function:

$$\mathcal{L}(\mathbf{y}, \hat{\mathbf{y}}) = \mathrm{CE}(\mathbf{y}, \hat{\mathbf{y}}) + \lambda \sum_{i,j,c} |\mathbf{A}_c^{ij}|. \tag{1}$$

Here, CE denotes the cross-entropy loss, and $\mathbf{y}$ represents the reference class labels. The sparsity of the evidence maps is controlled by the hyperparameter λ. The entire model is trained end-to-end using gradient descent.

3 Results

3.1 Datasets

We used two publicly available retinal fundus datasets, the Kaggle Diabetic Retinopathy (DR) [12] and the Age-Related Eye Disease Study (AREDS) [13]. The Kaggle DR dataset had 45,923 images from 28,984 subjects after applying a custom quality filtering with class distributions: 73% No DR, 15% Mild, 8% Moderate, 3% Severe, and 1% Proliferative DR. The AREDS dataset contained 34,079 images from 4,757 participants. AMD severity was grouped into six categories [2,13]: 49%, 19%, 14%, 3%, 12%, 1% for early, moderate, advanced intermediate, early late, active neovascular, and end-stage AMD.

Images were resized to 512×512, normalized, and augmented with cropping, flipping, color jitter, and rotation. Datasets were split into 75% training, 10% validation, and 15% test, keeping each participant's records in the same split. We evaluated our model's ability to localize DR-related lesions against ground-truth human annotations using the IDRiD dataset [22], which provides 81 fundus images with pixel-level labels for microaneurysms (MA), hemorrhages (HE), soft exudates (SE), and hard exudates (EX). This enabled the assessment of interpretability through localization performance.

Table 1. Classification performance on the test sets. Reported computational costs include: parameters (M), memory (MB), and average inference time (s).

	Computational Cost			AREDS AMD		Kaggle DR	
	Par.	**Mem.**	**Time**	**Acc.**	κ	**Acc.**	κ
ViT [11]	86,094	341	09.5 ± 0.1	$.76 \pm .03$	$.90 \pm .02$	$.81 \pm .02$	$.71 \pm .04$
Swin [19]	86,883	358	15.5 ± 1.1	$.78 \pm 0.2$	$\mathbf{.92 \pm .02}$	$.85 \pm .02$	$.81 \pm .03$
ResNet [14]	23,518	101	04.2 ± 0.5	$.78 \pm .03$	$.89 \pm .02$	$.85 \pm .02$	$.81 \pm .03$
BagNet [17]	16,271	193	15.1 ± 0.1	$.75 \pm .03$	$.88 \pm .02$	$.86 \pm .02$	$.83 \pm .03$
ResNet-FCL-SA	69,732	281	06.2 ± 0.2	$.78 \pm .03$	$.90 \pm .02$	$.86 \pm .02$	$.82 \pm .03$
BagNet-FCL-SA	62,501	306	27.3 ± 0.2	$.77 \pm .03$	$.89 \pm .02$	$.85 \pm .02$	$.83 \pm .03$
ResNet-Conv-SA	69,735	285	06.3 ± 0.6	$.78 \pm .03$	$.91 \pm .02$	$.85 \pm .02$	$.83 \pm .03$
BagNet-Conv-SA	62,913	310	27.3 ± 0.3	$.77 \pm .03$	$.90 \pm .02$	$\mathbf{.87 \pm .02}$	$\mathbf{.84 \pm .02}$
sResNet-Conv-SA	69,735	285	06.3 ± 0.6	$\mathbf{.79 \pm .02}$	$.90 \pm .02$	$.85 \pm .02$	$.80 \pm .03$
sBagNet-Conv-SA	62,913	310	27.3 ± 0.3	$.77 \pm .03$	$.91 \pm .02$	$.85 \pm .02$	$.81 \pm .03$

3.2 Self-explanaible Hybrid Models Achieved SOTA Performance

We first evaluated our models on multiclass DR detection and AMD severity classification. Using the ResNet-50 and BagNet-33 as backbone, our model incorporated dual-resolution convolutional self-attention (DR-Conv-SA) and a

GDFN module [31]. We set the reduction factor to $r = 2$, following [15], and applied max pooling. The window size was tuned ($w = 10$ for BagNet, $w = 8$ for ResNet), along with the regularization coefficient λ (Eq. 1), to balance classification accuracy and evidence map sparsity. Classification metrics are reported with 95% confidence interval (CI) lengths from bootstrapping, while inference time is reported with standard deviation (SD) over 1,000 runs on the same input.

We compared our sparse models to the dense counterparts ($\lambda = 0$), a variant using linear self-attention (SA) with fully connected layer (FCL) classifier, and several other baselines: ResNet-50, BagNet-33, ViT-32 (input size 384), and Swin Transformer (input size 384, patch size 4, window size 12). All models were initialized with pre-trained weights from ImageNet and trained with the same setup: data augmentation, cross-entropy loss, cosine learning rate schedule, and SGD optimizer (learning rate 10^{-4}, weight decay 5×10^{-4}) on an NVIDIA A40 GPU, using PyTorch, with model selection based on the best validation accuracy.

Our interpretable-by-design hybrid models achieved state-of-the-art performance on both tasks. The dense model with BagNet backbone yielded the best results for DR classification, while the model with the ResNet backbone achieved the highest Cohen's kappa (κ) for AMD severity classification (Table 1). Despite the sparsity penalty on the class activation map, the sparse models maintained competitive accuracy, with only a slight reduction in κ. Notably, for AMD detection, κ exceeded accuracy, likely due to misclassifications occurring predominantly between adjacent severity levels (Fig. 3d). Overall, the computational cost varies depending on the backbone but remains lower than that of ViTs.

3.3   Sparsity Constraints Enhance Class Evidence Maps

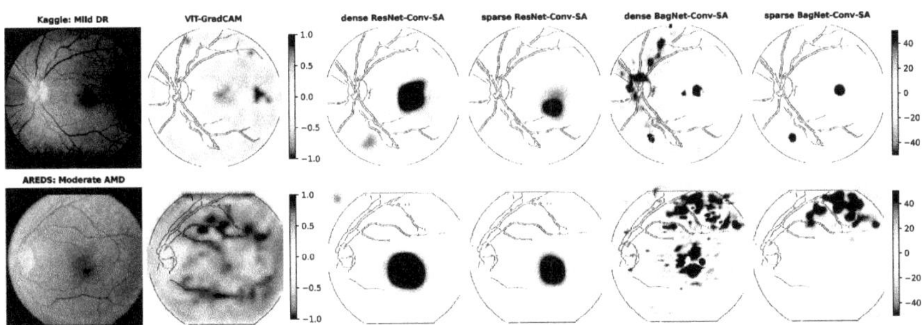

Fig. 2. **Examples explanations.** From left to right, heatmaps for the correctly predicted class. The first row shows an example (grade 1) from the Kaggle dataset, while the second row shows an example (grade 2) from the AREDS dataset.

We next compared evidence maps from our model to attribution maps generated with GradCAM [24] on the ViT baseline. As these were multiclass tasks, we only showed class evidence maps from the correctly predicted class. Our class evidence maps, obtained from the convolutional layer before average pooling, clearly highlighted input features relevant to the predicted class (Fig. 2). We noticed that GradCAM on ViT produced cluttered, hard-to-interpret heatmaps. In contrast, the hybrid ResNet-Transformer generated coarser heatmaps due to its large receptive field, while the hybrid BagNet-Transformer provided more localized explanations. The sparse models further refined this by producing sparser heatmaps, focusing decisions on smaller yet relevant retinal regions. For AMD severity classification, we observed that both the dense and sparse ResNet-Transformer models focus mainly on the macular region.

3.4 Evidence Maps Provide Faithful and Localized Explanations

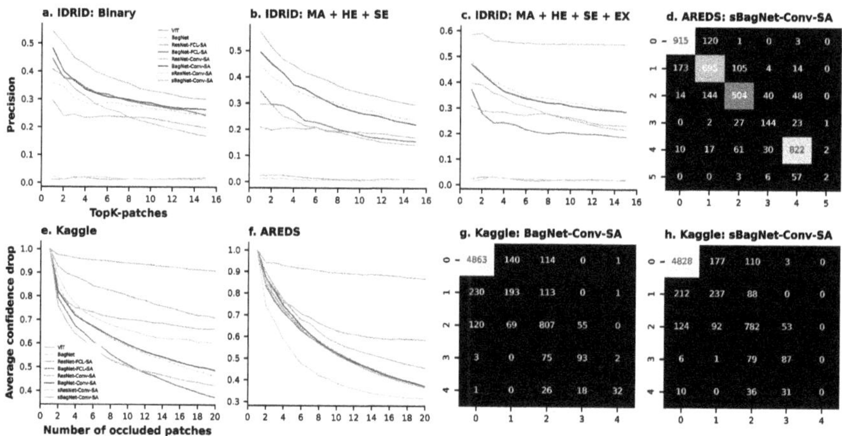

Fig. 3. Quantitative evaluation of heatmaps and confusion matrices. (**a-c**) Precision evaluation on IDRiD dataset. (**e,f**) Sensitivity analysis of different heatmaps for DR detection and AMD severity classification. (**d,g,h**) Confusion matrices of different models for DR detection and AMD severity classification on the test sets.

We quantitatively assessed the alignment of the explanations with clinical lesion-wise ground truth annotations by evaluating their precision in identifying DR lesions. Following the International Clinical Diabetic Retinopathy Scale [28], we evaluated three cases: (a) binary evaluation (Fig. 3a), averaging disease-class heatmaps and combining all lesion annotations; (b) severe DR (Fig. 3b), where MAs, HEs, and SEs were combined, and the precision was computed from the severe grade heatmap (c) proliferative DR (Fig. 3c), where all lesions were combined and precision was evaluated from the heatmap from the proliferative grade heatmap. Precision was measured as the proportion of positively activated regions containing lesions [17], using 33×33 non-overlapping patches to match BagNet's receptive field. For ViT and hybrid FCL models, GradCAM-generated heatmaps were used. Patches were extracted from positively activated regions. In all cases, the sparse BagNet-Transformer showed considerably higher precision than all other models and outperformed the base BagNet, suggesting that incorporating attention improved both classification and interpretability. The ResNet-Transformer with an explicit class-evidence layer performed worse, likely due to its larger receptive field producing coarser localizations (Fig. 2).

Subsequently, we additionally measured the faithfulness of the explanations by evaluating their ability to identify relevant regions for classification [30]. Using correctly classified test images, we progressively removed top-ranked patches highlighted in the heatmap and measured the resulting drop in class confidence. For DR detection, the sparse BagNet-Transformer performed best, while standard ViTs performed worst, followed by the ResNet-Transformer (Fig. 3e). In contrast, for AMD severity classification, the hybrid sparse ResNet outperformed the sparse BagNet-Transformer (Fig. 3f), likely due to the larger lesion sizes in AMD, which favor CNNs with larger receptive fields. Notably, this trend was consistent with classification results, where the ResNet backbone also excelled.

3.5   Our Model Enhances Interpretability for Multi-Class Tasks

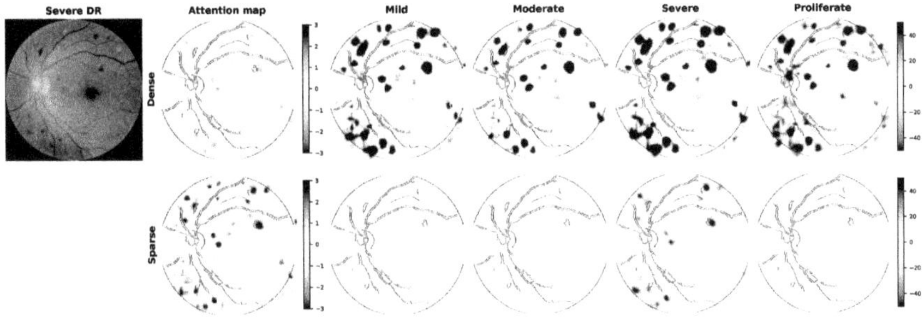

Fig. 4. Examples of multi-class explanations. Class-specific heatmaps for a Severe DR example from the Kaggle dataset. The first row displays the attention map and corresponding heatmaps from the dense hybrid model with the BagNet backbone, while the second row shows the attention map and heatmaps from its sparse version.

Finally, we visualized class-specific explanations for the dense and sparse BagNet-Transformer. For DR prediction on the Kaggle dataset, both models correctly classified the example (Fig. 4). For our hybrid model, heatmaps and class probability distributions were generated in a single forward pass, with the sparse model producing more focused and localized explanations aligned with both the predicted class and clinical ground truth. In contrast, post-hoc explanations required multiple forward passes, increasing the overall inference cost. In other classes, the sparse model showed almost no positive activations, unlike the dense model, which presented a mix of positive and negative evidence. Interestingly, we observed a strong correlation between attention maps and predicted evidence maps, particularly in the sparse model. This suggests that the model effectively captures long-range dependencies in an interpretable way.

4   Discussion and Conclusion

We introduced the first inherently interpretable hybrid CNN-transformer architecture for medical image classification[1], applied to DR detection and AMD severity classification from retinal fundus images. The approach is backbone-agnostic, allowing backbone selection to be guided by disease-specific prior. We evaluated the model with two CNN backbones: ResNet-50, which captured global spatial relationships relevant to AMD, and BagNet, which aggregates small local features important for DR detection. The latter is particularly noteworthy, as the SA mechanism helps overcome BagNet's limited receptive field. Conversely, ResNet's larger receptive field is better suited for AMD, which involves larger lesions. In both cases, SA enhances the model's focus on the most relevant features. Our transformer module employs dual-resolution convolutional SA to capture both global and fine-grained features while preserving strong local inductive biases. Unlike standard models with FCL classifiers, our model includes an explicit class evidence layer that produces spatial class-evidence heatmaps, enabling direct interpretability without post-hoc methods.

Interestingly, the interpretability-accuracy tradeoff was relatively small, challenging the myth of the accuracy-interpretability tradeoff in self-explainable models [23]. All evaluated models achieved comparable performance, with high balanced accuracy and κ. Notably, the sparse BagNet-Transformer produced the most informative explanations for DR detection, while the sparse ResNet-Transformer yielded the best explanations for AMD severity classification.

Preliminary experiments showed that multi-head SA increased training time without improving classification, while multi-scale resolution had limited impacts and further increased both the training and inference time—particularly with the BagNet backbone (Table 1), due to its larger feature maps and the resulting higher SA computation cost. Following [17], we also observed that higher sparsity often led to missed detection of late-stage DR, likely due to their underrepresentation in the training set (Fig. 4h). However, our hybrid architecture mitigated this issue more effectively, demonstrating robustness in low-data settings. Overall, our findings underscore hybrid CNN-Transformer models as a strong alternative to post-hoc ViT explanations, particularly for medical imaging.

Acknowledgments. This project was supported by the Hertie Foundation, the German Science Foundation (Excellence Cluster EXC 2064 "Machine Learning—New Perspectives for Science", project number 390727645). The authors thank the International Max Planck Research School for Intelligent Systems (IMPRS-IS) for supporting KD. PB is a member of the Else-Kröner-Kolleg "ClinBrAIn".

Disclosure of Interests. The authors declare no competing interests.

---

[1] Code at https://github.com/kdjoumessi/Self-Explainable-CNN-Transformer.

References

1. Abnar, S., Zuidema, W.: Quantifying attention flow in transformers. In: Proceedings of the 58th Annual Meeting of the Association for Computational Linguistics, pp. 4190–4197 (2020)
2. Al-Zamil, W.M., Yassin, S.A.: Recent developments in age-related macular degeneration: a review. Clin. Interv. Aging 1313–1330 (2017)
3. Azad, R., et al.: Advances in medical image analysis with vision transformers: a comprehensive review. Med. Image Anal. **91**, 103000 (2024)
4. Bach, S., Binder, A., Montavon, G., Klauschen, F., Müller, K.R., Samek, W.: On pixel-wise explanations for non-linear classifier decisions by layer-wise relevance propagation. PLoS ONE **10**(7), e0130140 (2015)
5. Bibal, A., et al.: Is attention explanation? An introduction to the debate. In: Proceedings of the 60th Annual Meeting of the Association for Computational Linguistics (Volume 1: Long Papers), pp. 3889–3900 (2022)
6. Brendel, W., Bethge, M.: Approximating CNNs with bag-of-local-features models works surprisingly well on imagenet. In: International Conference on Learning Representations (2019)
7. Chefer, H., Gur, S., Wolf, L.: Generic attention-model explainability for interpreting bi-modal and encoder-decoder transformers. In: Proceedings of the IEEE/CVF International Conference on Computer Vision, pp. 397–406 (2021)
8. Chefer, H., Gur, S., Wolf, L.: Transformer interpretability beyond attention visualization. In: Proceedings of the IEEE/CVF Conference on Computer Vision and Pattern Recognition, pp. 782–791 (2021)
9. Chen, C., Isa, N.A.M., Liu, X.: A review of convolutional neural network based methods for medical image classification. Comput. Biol. Med. **185**, 109507 (2025)
10. Djoumessi, K., Berens, P.: Soft-cam: making black box models self-explainable for high-stakes decisions. arXiv preprint arXiv:2505.17748 (2025)
11. Dosovitskiy, A., et al.: An image is worth 16×16 words: transformers for image recognition at scale. In: 9th International Conference on Learning Representations, ICLR (2021)
12. Dugas, E., Jared, J., Cukierski, W.: Diabetic retinopathy detection (2015). https://kaggle.com/competitions/diabetic-retinopathy-detection
13. Group, A.R.E.D.S.R.: The age-related eye disease study (areds): design implications areds report no. 1. Controlled Clin. Trials **20**(6), 573–600 (1999)
14. He, K., Zhang, X., Ren, S., Sun, J.: Deep residual learning for image recognition. In: Proceedings of the IEEE Conference on Computer Vision and Pattern Recognition, pp. 770–778 (2016)
15. Ilyas, Z., et al.: A hybrid CNN-transformer feature pyramid network for granular abdominal aortic calcification detection from DXA images. In: International Conference on Medical Image Computing and Computer-Assisted Intervention, pp. 14–25. Springer, Cham (2024)
16. Kashefi, R., Barekatain, L., Sabokrou, M., Aghaeipoor, F.: Explainability of vision transformers: a comprehensive review and new perspectives. arXiv preprint arXiv:2311.06786 (2023)
17. Kerol, D., et al.: Sparse activations for interpretable disease grading. In: Medical Imaging with Deep Learning (2023)
18. Kim, J.W., Khan, A.U., Banerjee, I.: Systematic review of hybrid vision transformer architectures for radiological image analysis. J. Imaging Inform. Med. 1–15 (2025)

19. Liu, Z., et al.: Swin transformer: hierarchical vision transformer using shifted windows. In: Proceedings of the IEEE/CVF International Conference on Computer Vision, pp. 10012–10022 (2021)
20. Maurício, J., Domingues, I., Bernardino, J.: Comparing vision transformers and convolutional neural networks for image classification: a literature review. Appl. Sci. **13**(9), 5521 (2023)
21. Nguyen, H.C., Lee, H., Kim, J.: Inspecting explainability of transformer models with additional statistical information. arXiv preprint arXiv:2311.11378 (2023)
22. Porwal, P., et al.: Indian diabetic retinopathy image dataset (idrid): a database for diabetic retinopathy screening research. Data **3**(3), 25 (2018)
23. Rudin, C.: Stop explaining black box machine learning models for high stakes decisions and use interpretable models instead. Nat. Mach. Intell. **1**(5), 206–215 (2019)
24. Selvaraju, R.R., Cogswell, M., Das, A., Vedantam, R., Parikh, D., Batra, D.: Gradcam: visual explanations from deep networks via gradient-based localization. In: Proceedings of the IEEE International Conference on Computer Vision, pp. 618–626 (2017)
25. Stassin, S., Corduant, V., Mahmoudi, S.A., Siebert, X.: Explainability and evaluation of vision transformers: an in-depth experimental study. Electronics **13**(1), 175 (2023)
26. Takahashi, S., et al.: Comparison of vision transformers and convolutional neural networks in medical image analysis: a systematic review. J. Med. Syst. **48**(1), 1–22 (2024)
27. Vaswani, A., et al.: Attention is all you need. In: Guyon, I., et al. (eds.) Advances in Neural Information Processing Systems, vol. 30 (2017)
28. Wilkinson, C.P., et al.: Proposed international clinical diabetic retinopathy and diabetic macular edema disease severity scales. Ophthalmology **110**(9), 1677–1682 (2003)
29. Wu, H., et al.: CVT: introducing convolutions to vision transformers. In: Proceedings of the IEEE/CVF International Conference on Computer Vision, pp. 22–31 (2021)
30. Yeh, C.K., Hsieh, C.Y., Suggala, A., Inouye, D.I., Ravikumar, P.K.: On the (in) fidelity and sensitivity of explanations. In: Advances in Neural Information Processing Systems, vol. 32 (2019)
31. Zamir, S.W., Arora, A., Khan, S., Hayat, M., Khan, F.S., Yang, M.H.: Restormer: efficient transformer for high-resolution image restoration. In: Proceedings of the IEEE/CVF Conference on Computer Vision and Pattern Recognition, pp. 5728–5739 (2022)

Robust Input Feature Attribution Maps
for Deep Neural Networks

Rachana Sathish and V. Rahul$^{(\boxtimes)}$

GE HealthCare, Bengaluru, India
`rahul.venkataramani@gehealthcare.com`

Abstract. Deep neural networks (DNNs) have shown remarkable performance in medical imaging tasks, yet their black-box nature hinders adoption in safety-critical clinical settings. Feature attribution maps offer a means of interpretability by highlighting input regions that influence model decisions. However, these maps are often noisy and sensitive to model parameters and input perturbations, limiting their reliability. We propose a robust attribution framework that utilize Test-Time Augmentation (TTA) to enhance the stability of feature attribution maps. Our method improves attribution consistency by aggregating attribution maps generated from multiple perturbed inputs, incorporating both stochastic noise and geometric transformations to capture a broader range of input variations. We experimentally demonstrate that the proposed method produces significantly more reliable results, with a 35% improvement on the Pointing Game metric, a standard for assessing attribution accuracy.

Keywords: Feature attribution · Explainability · Test-time augmentation

1 Introduction

Deep neural networks (DNNs) have become the choice of tool in developing artificial intelligence (AI) based solutions for medical imaging, achieving state-of-the-art performance across a wide range of tasks, including classification, segmentation and reconstruction. However, in high-stakes domains such as healthcare, the adoption of DNNs is often hindered by their opaque, black-box nature. Clinicians and decision-makers require not only accurate predictions but also transparent and interpretable justifications to ensure patient safety and build trust in AI-assisted workflows.

To address this challenge, a variety of interpretability techniques have been proposed. Among them, feature attribution maps, also known as saliency maps, have gained popularity for their intuitive visual explanations. Gradient-based attribution methods, such as GradCAM [10], that compute the gradient of output or activation with respect to the input have proven to be unreliable due to gradient saturation [11] in back-propagation. Integrated Gradients (IG) [16]

© The Author(s), under exclusive license to Springer Nature Switzerland AG 2026

M. Reyes et al. (Eds.): iMIMIC 2025, LNCS 16464, pp. 117–126, 2026.
https://doi.org/10.1007/978-3-032-17611-0_12

addresses this limitation by computing feature attributions as the path integral of gradients along a straight-line interpolation between a baseline $\mathbf{B}$, which serves as a counterfactual reference, and the actual input $\mathbf{I}$. The reliability of the computed attribution map in IG depends on several factors, including the number of steps used to approximate the path integral, the choice of baseline $\mathbf{B}$, and the specific path taken from $\mathbf{B}$ to the input $\mathbf{I}$. Figure 1 presents a comparison of attribution map generated by the proposed method and IG, where IG map exhibit noticeable noise that can obscure salient features.

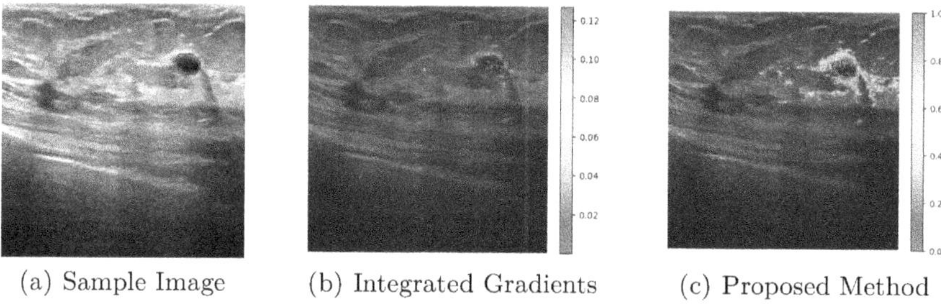

(a) Sample Image   (b) Integrated Gradients   (c) Proposed Method

Fig. 1. Figure shows (a) sample breast ultrasound image with the contour of the tumor marked in red and attribution maps generated using (b) IG and (c) proposed method. (Color figure online)

The choice of number of steps M from the baseline image $\mathbf{B}$ and the input image $\mathbf{I}$ for computing IG is defined by the computational budget. In this work, we propose a novel approach to enhance the robustness of feature attribution maps by leveraging Test-Time Augmentation (TTA) which is traditionally used to improve prediction accuracy by averaging outputs over multiple augmented versions of the input. The proposed method operates under the same computational budget which is distributed across the augmented samples of the input image as shown in Fig. 2. The number of steps from $\mathbf{B}$ to the augmented image $\mathbf{I}_i$ is reduced by a factor N, i.e., $M' = \frac{M}{N}$, where N is the number of augmentation transformations.

Our key contributions are as follows:

– We introduce a TTA-based framework for generating robust feature attribution maps.
– We demonstrate that our method significantly reduces attribution noise and improves consistency by aggregating across multiple perturbations.
– We validate our approach on a benchmark dataset of breast ultrasound dataset by showing significant improvement over baseline attribution methods.

2   Related Works

Existing methods for input feature attribution are broadly categorized into *intrinsic* approaches, which design interpretable models from the ground up, and *post-hoc* approaches, which explain the behavior of already-trained models. This paper proposes a robust post-hoc approach that is independent of the model architecture.

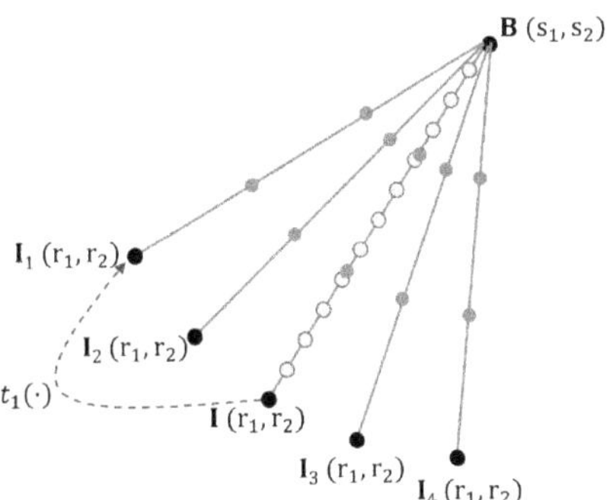

Fig. 2. Graphical illustration demonstrating the utilization of computational budget in IG vs proposed method. The green lines represent the straight line path along which the gradients are accumulated in IG with the yellow dots denoting the steps along the path. Similarly, the red dots denote the steps along the path for the augmented images I_i used in the proposed method. (Color figure online)

2.1   Interpretability Methods

Among post-hoc interpretability techniques, feature attribution methods are widely used due to their intuitive visual explanations. These methods assign importance scores to input features based on their influence on the model's output. Gradient-based methods such as Saliency Maps [12], Guided Backpropagation [15], and Grad-CAM [10] compute derivatives of the output with respect to the input to highlight influential regions. Integrated Gradients [16] address gradient saturation by accumulating gradients along a path from a baseline input to the actual input. Perturbation-based methods like Occlusion, LIME [9], and SHAP [7] estimate feature importance by observing changes in the output when parts of the input are modified or masked.

Despite their utility, these methods often suffer from instability. Small perturbations in the input or variations in model parameters can lead to significantly different attribution maps, raising concerns about their reliability in critical applications.

2.2 Robustness in Interpretability

To address the instability of attribution methods, several techniques have been proposed. SmoothGrad [14], for instance, averages attribution maps over multiple noisy versions of the input to reduce visual noise. VarGrad [1] extends this idea by analyzing the variance of attributions across perturbations. A similar extension [5] has also been proposed for IG using Taylor's theorem. IG^2 [17] proposed to improve IG by accumulating the gradients along a generated path instead of the straight line path. Additionally, the zero baseline is replaced with a set of reference images. Ensemble-based approaches aggregate explanations from multiple models or model checkpoints to improve consistency [6]. While these methods offer improvements, they often require additional computational resources or are limited in their generalizability.

2.3 Test-Time Augmentation

Test-Time Augmentation (TTA) is a well-established technique in computer vision, primarily used to enhance prediction [8,13] accuracy by averaging outputs over multiple augmented versions of the input. TTA has been successfully applied in classification, segmentation, and detection tasks. However, its potential for improving interpretability has been largely unexplored.

While existing methods attempt to mitigate the noise in attribution maps, they often fall short in balancing robustness, simplicity, and computational efficiency. Our work addresses this gap by introducing a TTA-based framework that aggregates attribution maps across augmented inputs, offering a principled and effective solution for generating robust explanations.

3 Methodology

3.1 Problem Definition

Let $\mathbf{I} \in \mathbb{R}^{H \times W \times C}$ denote an input image, where H, W, and C represent the height, width, and number of channels, respectively. Let $f_\theta(\cdot)$ be a DNN parameterized by θ, which maps the input image to a prediction $p = f_\theta(\mathbf{I})$. Our goal is to generate a robust feature attribution map $\overline{M} \in \mathbb{R}^{H \times W}$ that highlights the regions in $\mathbf{I}$ most influential in the model's decision. Although our approach is compatible with a wide range of feature attribution methods, we illustrate its application in the context of Integrated Gradients in the subsequent sections, given that many other attribution techniques can be viewed as extensions or variations of this foundational method.

3.2 Attribution Computation Using Integrated Gradients

Given a trained model f_θ, the IG attribution for image $\mathbf{I}$ is computed as,

$$IG(\mathbf{I}) = (\mathbf{I} - \mathbf{B}) \int_{\alpha=0}^{1} \frac{\partial f_\theta\left(\mathbf{I} + \alpha(\mathbf{I} - \mathbf{B})\right)}{\partial \mathbf{I}} \partial\alpha \tag{1}$$

where $\mathbf{B}$ is the baseline image and $\alpha \in [0, 1]$ is the scaling parameter for the input.

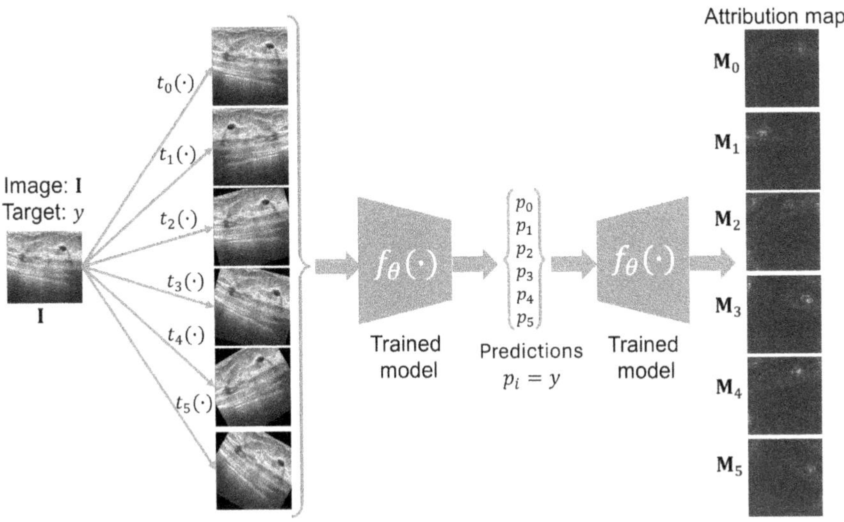

Fig. 3. Overview of the proposed method. Test-time augmentation is performed on the input to a trained model to generate multiple feature attribution maps. These maps are aggregated after applying the inverse transformation corresponding to the augmentation.

3.3   Test-Time Augmentation Framework

To improve the robustness of the attribution map, we introduce a Test-Time Augmentation (TTA) strategy. Let $\mathcal{T} = \{t_1, t_2, \ldots, t_N\}$ be a set of transformations (e.g., flips, rotations). For each transformation $t_n \in \mathcal{T}$, we compute the augmented input as $\mathbf{I}_n = t_n(\mathbf{I})$. This is illustrated in Fig. 3. Using IG, attribution map $\mathbf{M}_n$ for each augmented image is computed as in Eq. 1.

Since each $\mathbf{M}_n$ corresponds to a transformed version of the input, we apply the inverse transformation t_n^{-1} as shown in Fig. 4 to align it with the original input space before aggregating them. The final attribution map is therefore computed as,

$$\overline{\mathbf{M}} = \frac{1}{N} \sum_{n=1}^{N} b\left(t^{-1}\left((\mathbf{I}_n - \mathbf{B}) \int_{\alpha=0}^{1} \frac{\partial f_\theta\left(\mathbf{I}_n + \alpha(\mathbf{I}_n - \mathbf{B})\right)}{\partial \mathbf{I}_n} \partial\alpha \right) \right) \tag{2}$$

where $b(\cdot)$ converts the continuous valued attribution map to a binary map by retaining only the top-$k\%$ of the attribution values.

4   Experiments and Results

4.1   Training Classification Model

The proposed method is experimentally validated on the task of classifying breast ultrasound images into three classes - normal, benign and malignant. This is

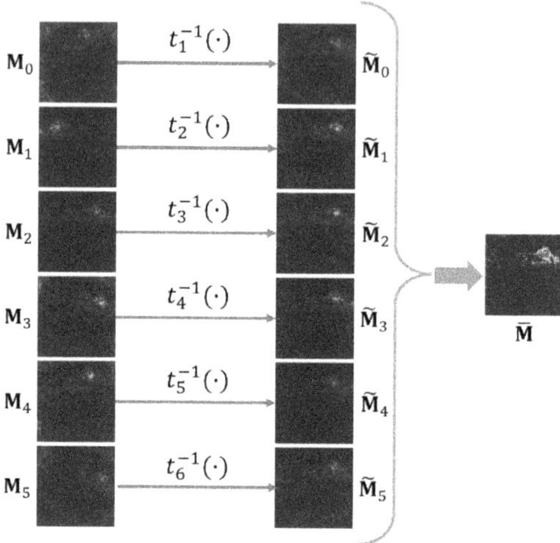

Fig. 4. Aggregation of aligned attribution maps.

demonstrated on the BUSI dataset [2] which includes segmentation masks for tumors in addition to the class label. We followed the 60/20/20 strategy for creating training, validation and test split from the dataset where samples belonging to each class were randomly assigned to one of the splits while preserving the class distribution across the splits.

We trained a transformer-based classification model with ViT-B [4] as the backbone. The model was trained for 100 epochs with a learning rate of 10^{-4} with early stopping to prevent over-fitting. The trained classification model achieved an accuracy of 92.41% on the test split.

4.2 Evaluation of Attribution Maps

In our implementation of the proposed method, we applied five augmentations including horizontal flip and rotation by $10, -10, 20, -20°$. The framework is model-agnostic and can be applied to any attribution method. The proposed TTA-based technique is compared with the following baselines:

1. **Grad-CAM** [10] computes a coarse localization map highlighting the important regions in the input by using the gradient of the target class in the pre-final layer of the network.
2. **Integrated Gradients** (IG) [16] attributes the output of a model f_θ to the input features by integrating the gradients of f_θ along a straight line path from a baseline to the actual input. In practical implementation, an approximation of this is obtained by performing a summation of the gradients computed at S number of steps along the path. As stated in Sect. 1, S is constrained by the computational budget.

3. **SmoothGrad** [14] reduces the noise in the attribution map by averaging the gradients of multiple noisy versions of the input. In our experiments, IG with SmoothGrad is used as one of the baseline methods.
4. **IG^2** [17] addresses the issue of noisy attributions and zero baseline by combining the gradients of the explicand's class and the counterfactual class.

We use $S = 100$ in IG and set that as the fixed computational budget. Following this, 5 random samples were used per image in SmoothGrad with $S = 20$ for each image. Similarly, we use $S = 20$ for the proposed TTA-based method which uses 5 augmentation transformations. In the case of IG^2, 5 reference images of a different class than the target were used with $S = 20$.

Evaluation of the baselines and the proposed method is done using the metrics reported in [3]. Since the pizel-wise annotations for the tumors were available in the dataset, we computed Intersection over Union (IoU) between the binarized attribution maps and the segmentation masks instead of using bounding boxes. The second metric, success of pointing game, reported in [3] measures the percentage of samples for which the maximum value of the attribution map lies within the region of interest (tumor). We note that this metric is particularly important for clinicians to highlight the most important region for a given sample. The quantitative results are presented in Table 1. Visualization of the attribution maps for the baseline approaches and the proposed method is presented in Fig. 5.

Table 1. Quantitative Evaluation of Attribution Maps

Method	TTA (Proposed)	#steps	IoU		Pointing Game	
			Benign	Malignant	Benign	Malignant
GradCAM	✗	-	0.114 ± 0.142	0.042 ± 0.041	42.50%	55.88%
	✓	-	0.225 ± 0.213	0.134 ± 0.128	68.75%	91.18%
SmoothGrad-IG	✗	20	0.227 ± 0.142	0.234 ± 0.133	51.25%	58.28%
IG^2	✗	20	0.105 ± 0.051	0.123 ± 0.070	99.34%	97.18%
IG	✗	100	0.229 ± 0.144	0.233 ± 0.131	51.25%	61.76%
	✓	20	0.270 ± 0.234	0.123 ± 0.099	91.25%	97.06%

5   Discussion

Our proposed test-time augmentation (TTA) based attribution framework builds on existing gradient-based methods. Instead of relying on a single input or a fixed baseline, our approach aggregates attributions over a set of perturbed inputs sampled from a local neighborhood around the original image. While conceptually related to SmoothGrad, which averages gradients over noisy perturbations, our method generalizes the notion of locality to include geometric transformations that preserve task-relevant semantics. This broader neighborhood allows

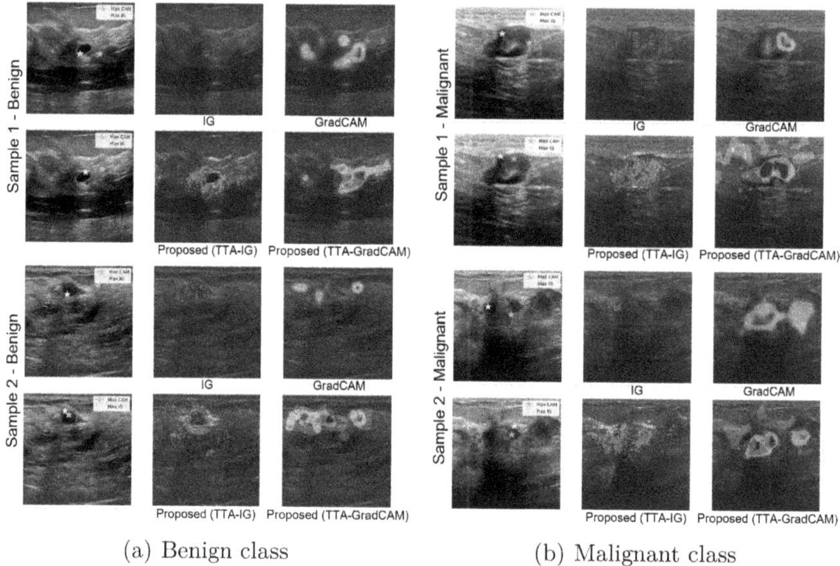

(a) Benign class   (b) Malignant class

Fig. 5. Figure shows qualitative results for a few sample images from the dataset. The contour of the lesion is drawn in red over the image and the attribution maps. The yellow star overlayed on the image indicates the location with maximum value for the IG attribution map and the one in green corresponds to GradCAM. (Color figure online)

the model to identify features that are consistently salient across a range of plausible, task-invariant variations.

Empirical results show that this TTA-based strategy produces more stable and accurate attributions. This is particularly valuable in domains where traditional baselines are ill-defined or ineffective. By leveraging consistency under structured perturbations, the method enhances interpretability without requiring domain-specific counterfactual generation. The proposed TTA-based approach enhances the baseline attribution methods, Grad-CAM and Integrated Gradients (IG), in terms of both Intersection over Union (IoU) and the Pointing Game metric. Since SmoothGrad inherently incorporates input perturbations to generate attribution maps, the TTA-based extension is not applied to it. A similar rationale applies to IG^2, which utilizes multiple reference images. Although the proposed method exhibits slightly lower performance than IG^2 in the Pointing Game, it is important to note that IG^2 is computationally more intensive due to the additional overhead of generating the gradient path (GradPath) and the baseline (GradCF) through a separate optimization process.

6   Conclusion

In this work, we addressed the challenge of generating robust and reliable feature attribution maps for deep neural networks, particularly in safety-critical domains such as healthcare. While existing attribution methods provide valuable insights into model behavior, their sensitivity to input perturbations and model parameters often undermines their trustworthiness. We proposed a novel framework that leverages Test-Time Augmentation (TTA) to enhance the robustness of attribution maps. By aggregating explanations across multiple augmented inputs, our method reduces noise and improves consistency, offering more stable and interpretable visualizations. This approach is simple, model-agnostic, and can be integrated with a variety of existing attribution techniques. Experimental results demonstrate that our TTA-based method significantly improves the reliability of feature attributions, making it a promising tool for real-world deployment where interpretability and trust are paramount. Future work will explore adaptive augmentation strategies and extend the framework to other interpretability paradigms beyond feature attribution.

References

1. Adebayo, J., Gilmer, J., Muelly, M., Goodfellow, I., Hardt, M., Kim, B.: Sanity checks for saliency maps. In: Advances in Neural Information Processing Systems, pp. 9505–9515 (2018)
2. Al-Dhabyani, W., Gomaa, M., Khaled, H., Fahmy, A.: Dataset of breast ultrasound images. Data Brief **28**, 104863 (2020)
3. Chung, M., Won, J.B., Kim, G., Kim, Y., Ozbulak, U.: Evaluating visual explanations of attention maps for transformer-based medical imaging. In: International Conference on Medical Image Computing and Computer Assisted Intervention, pp. 110–120 (2024)
4. Dosovitskiy, A., et al.: An image is worth 16×16 words: transformers for image recognition at scale. In: International Conference on Learning Representations
5. Goh, G.S., Lapuschkin, S., Weber, L., Samek, W., Binder, A.: Understanding integrated gradients with smoothtaylor for deep neural network attribution. In: International Conference on Pattern Recognition, pp. 4949–4956 (2021)
6. Kadir, M.A., Addluri, G., Sonntag, D.: Harmonizing feature attributions across deep learning architectures: enhancing interpretability and consistency. In: German Conference on Artificial Intelligence, pp. 90–97 (2023)
7. Lundberg, S.M., Lee, S.I.: A unified approach to interpreting model predictions. In: Advances in Neural Information Processing Systems, pp. 4765–4774 (2017)
8. Moshkov, N., Mathe, B., Kertesz-Farkas, A., Hollandi, R., Horvath, P.: Test-time augmentation for deep learning-based cell segmentation on microscopy images. Sci. Rep. **10**(1), 5068 (2020)
9. Ribeiro, M.T., Singh, S., Guestrin, C.: "Why should I trust you?": explaining the predictions of any classifier. In: International Conference on Knowledge Discovery and Data Mining, pp. 1135–1144 (2016)
10. Selvaraju, R.R., Cogswell, M., Das, A., Vedantam, R., Parikh, D., Batra, D.: Gradcam: visual explanations from deep networks via gradient-based localization. In: International Conference on Computer Vision, pp. 618–626 (2017)
11. Shrikumar, A., Greenside, P., Kundaje, A.: Learning important features through propagating activation differences. In: International Conference on Machine Learning, pp. 3145–3153 (2017)

12. Simonyan, K., Vedaldi, A., Zisserman, A.: Deep inside convolutional networks: visualising image classification models and saliency maps. In: International Conference on Learning Representations (2014)
13. Simonyan, K., Zisserman, A.: Very deep convolutional networks for large-scale image recognition. arXiv preprint arXiv:1409.1556 (2014)
14. Smilkov, D., Thorat, N., Kim, B., Viégas, F., Wattenberg, M.: Smoothgrad: removing noise by adding noise. arXiv preprint arXiv:1706.03825 (2017)
15. Springenberg, J., Dosovitskiy, A., Brox, T., Riedmiller, M.: Striving for simplicity: the all convolutional net. In: International Conference on Learning Representations (Workshop) (2015)
16. Sundararajan, M., Taly, A., Yan, Q.: Axiomatic attribution for deep networks. In: International Conference on Machine Learning, pp. 3319–3328 (2017)
17. Zhuo, Y., Ge, Z.: Ig2: integrated gradient on iterative gradient path for feature attribution. IEEE Trans. Pattern Anal. Mach. Intell. (2024)

From Explainable to Explained AI: Ideas for Falsifying and Quantifying Explanations

Yoni Schirris[1(✉)], Eric Marcus[1], Jonas Teuwen[1], Hugo Horlings[1], and Efstratios Gavves[2]

[1] Netherlands Cancer Institute, Amsterdam, The Netherlands
yschirris@gmail.com
[2] University of Amsterdam, Amsterdam, The Netherlands

Abstract. Explaining deep learning models is essential for clinical integration of medical image analysis systems. A good explanation highlights if a model depends on spurious features that undermines generalization and harms a subset of patients or, conversely, may present novel biological insights. Although techniques like GradCAM can identify influential features, they are measurement tools that do not themselves form an explanation. We propose a humanmachine-VLM interaction system tailored to explaining classifiers in computational pathology, including multi-instance learning for whole-slide images. Our proof of concept comprises (1) an AI-integrated slide viewer to run sliding-window experiments to test claims of an explanation, and (2) quantification of an explanation's predictiveness using general-purpose vision-language models. The results demonstrate that this allows us to qualitatively test claims of explanations and can quantifiably distinguish competing explanations. This offers a practical path from explainable AI to explained AI in digital pathology and beyond. Code and prompts are available at https://github.com/nki-ai/x2x.

Keywords: Explainable AI · Computational Pathology · VLM

1 Introduction

The application of deep learning (DL) in medical image analysis has become ubiquitous across tasks and domains [2,17] and is increasingly being implemented in the clinic. However, the black-box nature of DL often leaves these models unexplained. Most DL models are correlational learners, often leading to shortcut learning [8]. The model then uses spurious features to make predictions that may lead to unwanted outcomes if implemented in the clinic. Examples are available for the predictions of the response to chemotherapy from breast magnetic resonance images [4,19] where the model remembered patient anatomy; in survival prediction from multimodal data including computational pathology [6,10] where medical center patterns themselves were predictive of survival; for

© The Author(s), under exclusive license to Springer Nature Switzerland AG 2026
M. Reyes et al. (Eds.): iMIMIC 2025, LNCS 16464, pp. 127–136, 2026.
https://doi.org/10.1007/978-3-032-17611-0_13

COVID-19 prediction from CT images [7] where the model focused on technical artifacts on a scan; and medical center memorization by pathology foundation models [14]. These examples show that high-performance metrics on a validation dataset are not proof of a good model that has learned the underlying biology and causal structure of the task. We believe that providing an explanation of what the model does, rigorously testing, and iterating upon the explanation can help prevent such outcomes and guide researchers to overcome these issues during model development, and is essential for the clinical applicability of these models (Fig. 1).

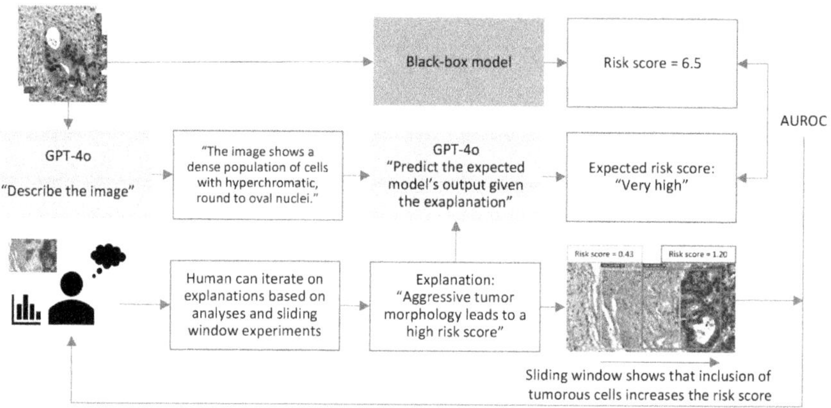

Fig. 1. Overview of our proposed evaluation framework for explanations. Sliding window experiments test the change in the black-box model output logit as specific features are included in the field of view, to test if the explanation could predict this. To quantitatively test if the hypothesized black-box explanation can accurately predict the output of the black-box model, the VLM follows an explanation to score a patch, which is compared to the black-box model output logit.

There is an increasing research interest in explainable AI (XAI) to explain how black-box models work in medical imaging [3,22]. For example, saliency methods like GradCAM [21] highlight the image's most influential parts for the current prediction. These visualizations are often called "the explanation" [3]. However, this visualization is only a measurement, which itself does not form an explanation. For example, the microscopic image of a striped jellybean inside a cell (mitochondria) told the viewer in the 19th century nothing of its role in energy production or its intricate evolutionary origins. In medical image analysis, another standard practice is to perform post-hoc correlative analyses of the model's output with clinical variables on the validation set [24]. Such studies investigate (in)dependence with (un)wanted variables and, combined with other XAI measurements, are used to generate an initial explanation of what the model might do. However, such an explanation is rarely tested further, as would be standard practice in the sciences. One reason for this omission is the complex collaboration required between highly trained medical and AI experts (e.g., [18] tests the biological validity of the explanation of [24]). We believe that another important reason is that *the generation and validation of good explanations of deep learning models is an open problem, lacking conceptual frameworks, practical guidelines, and toolkits.*

Problem : The absence of ideas and methods for the falsification and quantification of explanations for AI models limits the individual researcher to confidently provide good explanations to discover and prevent unwanted actions by the model.

To this end, we present a prototype for human-in-the-loop testing of components of a black box model's explanation in Sect. 4 and automated quantitative evaluation of such an explanation using a general-purpose vision-language model (VLM) in Sect. 5. Although this work focuses on a risk prediction model for colorectal cancer in computational pathology as described in Sect. 2, these ideas are transferrable to other tasks and domains.

Before we can investigate tools to challenge and quantify the predictive power of explanations, however, we should agree on what an "explanation" is.

2   What Is an Explanation?

We define an *explanation* of a DL model as a hypothesized natural language description of what the model does, such that it is predictive of the model's output for samples that fall within the distribution that the explanation is hypothesized for.

Imagine a cow-camel classifier [1] where the training set consists of cows in meadows and camels in deserts. Naively, we think the model distinguishes a cow from a camel independently of the background. It might do so in an internal validation dataset but, to our surprise, fails in an external test set - a *problem*. Further investigation reveals that the model acts independently of the animal, and is a good meadow-desert classifier - *an explanation*. This explanation, in some sense, solves the problem: we now understand *why* the model failed on the test dataset. We can use this explanation to design augmentations or generate a new dataset to train a better camel-cow classifier. An explanation should adhere to some standards to be helpful, though.

Good Explanations. Following the distinction between good and bad explanations by [16], an explanation is considered *good* if it adheres to three criteria. Firstly, (1) the explanation should be criticizable. In the case of the camel-cow classifier, we *can* experimentally test the explanation with images of animals in meadows and deserts and view the model's output. Additionally, (2) the explanation should be hard to vary. For the cow-camel classifier, for example, *we can not change* "meadow" to "airplane" while preserving the explanation's predictive power. Finally, (3) the explanation should be non-authoritarian. In the case of computational pathology, *it should not matter if a famous pathologist, PhD student in machine learning, or a VLM generates the explanation.* For instance,

"The model is a perfect survival predictor because famous pathologist X developed it" is a bad explanation because it does not explain anything about the reality of how the model functions. In this case, "the model provides a higher score for more aggressive tumor morphologies" is a better explanation.

Given a good explanation, we can start testing its predictive power: Can it accurately predict the output of the black-box model given an image?

3   An Initial Explanation of Prognosis Prediction

Now that we have a concept of explanations, we put it into practice. To do so, we test an open-source colorectal cancer risk prediction model by [13] for two reasons. Firstly, following posthoc correlative analyses, the authors forge an explanation that reaches similar conclusions to previous work [24], which we can challenge. Secondly, survival prediction is a complex task and may provide novel biological insights when presenting a good explanation, as shown by [18].

The model (which we term MIL) is a feature-extraction and multiple instance learning pipeline [12,20] that takes 224×224 px patches of the tissue from a hematoxylin and eosin stained whole-slide image (WSI) at 1.14 mpp, performs stain normalization [15], extracts features using RetCCL [23], and classifies an attention-weighted mean of all feature vectors to provide a WSI-level risk logit. See [13] for details. Here we will focus on why the model classifies a single patch as high- or low-risk, which is the elementary component of a WSI-level prediction.

Given the analytical results from [13], we conjecture an explanation: *"MIL assigns higher scores to patches exhibiting: (1) Poorly differentiated or highly proliferative tumor regions, often with epithelialmesenchymal transition. (2) Invasion into or infiltration of surrounding adipose tissue. (3) Morphological features indicative of an aggressive tumor-stroma interface. Conversely, lower scores are associated with better differentiation and organized immune infiltration."*. This explanation aligns with our criteria for good explanations from the last section: It is criticizable, hard-to-vary, and proposed due to evidence from analyses independent of who performed them.

Now the core problem that we wish to tackle arises: How do we criticize this initial explanation? For example, does MIL indeed predict a lower risk score for the same patch with aggressive tumor morphology if it would have contained immune infiltrate? The transition of morphologies in a WSI provides us a way to perform informal "interventional" experiments using sliding windows.

4   Testing and Falsifying Explanation Components With Intuitive Sliding-Window Experiments

The idea here is to find examples that disagree with, hence falsify, the provided explanation, similar to finding a black swan that falsifies the statement that all swans are white.

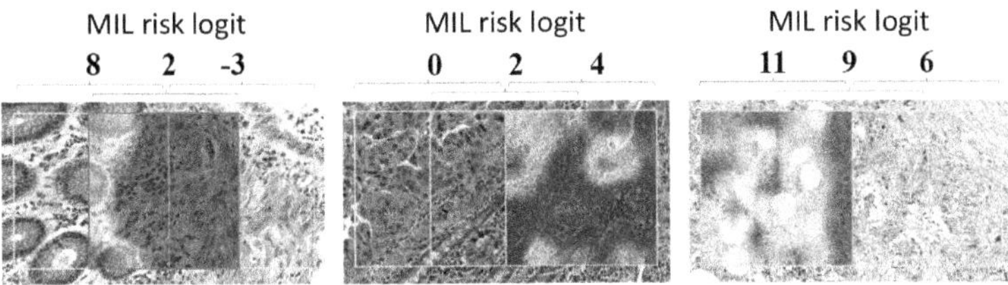

Fig. 2. Examples of sliding window experiments with an AI-integrated slide viewer. The risk logit as predicted by the MIL model is displayed on top of each patch. The heatmap displays a GradCAM measurement. Left: The right patch receives a low risk logit score, and as glandular features are included to the left, the risk logit increases. Middle: The MIL risk logit increases as lymphocytes are included to the right. Right: As we move to the left and include adipocytes that intermingle with the tumor cells in the right patch, this increases the MIL risk logit further.

We perform sliding window experiments to challenge components of an explanation on overlapping patches from WSI that preserve the original morphology while selectively including or excluding features of interest contained in the explanation. Using our explanation, these experiments assess whether the controlled modifications result in predictable changes in the MIL output. To perform the experiments effectively, we should be able to quickly and intuitively find and select such patches. To this end, we develop a WSI-viewer and integrate the model to perform patch extraction, normalization, inference, and GradCAM analysis on-the-fly. This allows the user to skim over the WSI to find features of interest, especially regions with a gradient of features to capture the change of the MIL predictions as various features are included. In Fig. 2, we test three components of the explanation: (1) the risk decrease of healthy glands, (2) the risk decrease of lymphocytic infiltration, and (3) the risk increase of tumor cells intermingling with adipocytes. We choose a location in the WSI that contains the histological morphology of the explanation's component we challenge, such that slightly moving a patch would include the other feature, which is done for three overlapping patches.

Healthy Glands: On the left of Fig. 2, the black-box risk score increases as normal glandular tissue is included in stromal tissue, confirmed by the GradCAM's highlighting of the glands. This contrasts the explanation in Sect. 2, which states that well-differentiated tissue leads to a lower risk score. This is not incidental: We often find that glandular structures receive very high risk scores. Although a resection of a tumor may rarely have healthy glands which are filtered out with the attention mechanism of MIL, this may pose a problem when a slide has a low tumor-to-tissue ratio.

Lymphocytic Infiltration: In the middle of Fig. 2, the inclusion of lymphocytic infiltration to an area with relatively organized and heterogeneous solid tumor tissue increases the risk score, confirmed by the GradCAM's focus on both tumor and lymphocytes. This contradicts the explanation, stating that immune response decreases the risk score. During the sliding window experiments the authors noticed that in many other WSI locations regions with lymphocytes do receive a low risk score, but this is not consistently so.

Adipocyte Infiltration: On the right of Fig. 2, we transition from a region of unorganized, pleomorphic tumor cells towards the left, where adipocytes intermingle with the tumor cells. The explanation states that the score would go up, which is confirmed as we include more and more adipocytes. The Grad-CAM highlights cells that border the adipocytes, suggesting that the proximity of tumor cells to adipocytes increases the score. These results are highly consistent.

To conclude, the existing explanation is already predictive, but these observations hint at incompleteness of the explanation as MIL's output logits do not always predictively change according to the explanation. The presented tool allows researchers to intuitively perform sliding-window experiments to test components of the explanation. Through refutation of existing components or addition of new insights, improved explanations of the model's functioning are attainable.

If a refined explanation is provided after these experiments, however, a new problem arises: How do we test if this refined explanation is more predictive than the original explanation on a larger dataset? Can we systematically quantify this?

5   Quantitatively Comparing Competing Explanations

Here, the idea is to investigate if we can automatically quantify the predictiveness of an explanation without human involvement to omit the requirement of pathologist annotations.

We test whether an image description combined with an explanation allows a general-purpose VLM to estimate ($\hat{y}$) the output logit of MIL ($y \in \mathbb{R}$). We use the GPT-4o API to provide a description of the histological features in a patch. Next, we input the image description and the putative explanation, and prompt GPT-4o to predict one of four classes:

$$\hat{y} \in \{\texttt{high, medium-high, medium-low, low}\}.$$

For all metric calculations, the four-way GPT-4o predictions are binarized (any high versus any low).

The test set X consists of 132 manually selected patches from 15 randomly samples WSIs from TCGA-CRC available from the GDC repository.

Since the explanations mostly distinguish between features that provide very high and very low scores and does not include an explanation for all possible morphologies present in a WSI, we expect it to be insufficient to distinguish between patches with similar scores close to the decision boundary. To this end, we define subsets $X_j = \{x_i \in X : |y_i| > j\}$, where samples with small MIL outputs (i.e., near the median score, the decision boundary) are removed.

We compute the bootstrapped AUROC for X (n=132, 55% high MIL risk logit), X_1 (n=81, 54% high MIL risk logit), X_3 (n=29, 72% high MIL risk logit), i.e., the subsets of the dataset with extreme predictions of MIL. We expect

GPT-4o with a poor explanation to fail on all these subsets, while a predictive explanation would achieve better performance on the easier datasets with highly diverging MIL risk logits.

Table 1. Explanations tested in our experiments.

Explicitly Poor Explanations

(1) cow_camel: The presence of a camel provides a high score, while the presence of a cow provides a low score. Absence of either leads to a medium-high or medium-low score.

(2) tumor_lymphocyte_inverse: Presence of lymphocytes leads to a higher score. Presence of tumor cells leads to a lower score.

Expected Predictive Explanations

(3) tumor_lymphocyte: Presence of lymphocytes leads to a lower score. Presence of tumor cells leads to a higher score.

(4) detailed_analyses: The explanation as described in Sect. 3.

To see if our method with the VLM and metric calculations is consistent and provides expected outputs, we want to see if an explanation that has nothing to do with pathology (cow camel) achieves a random performance. Additionally, to see if the VLM properly follows the explanation, we provide it with two inversed explanations (tumor lymphocyte (inverse)) to see if the resulting AUROC is inversed. Next, to see if these metrics can distinguish between two competing explanations that are expected to be predictive, we compare a very coarse explanation (tumor lymphocyte) with the explanation derived from detailed analyses (Sect. 3). These explanations are shown in Table 1 (Table 2).

Table 2. Bootstrapped AUROC using varying explanations on subsets of the dataset with increasingly divergent MIL risk logits

Explanation	Data Subset		
	X	X_1	X_3
cow camel	0.55 ± 0.05	0.55 ± 0.06	0.60 ± 0.11
tumor lymphocyte inverse	0.38 ± 0.06	0.31 ± 0.07	0.11 ± 0.07
tumor lymphocyte	0.62 ± 0.05	0.67 ± 0.07	0.77 ± 0.11
detailed analysis	0.62 ± 0.05	0.68 ± 0.06	0.79 ± 0.09

Explicitly Poor Explanations: Using the cow camel explanation, GPT-4o cannot find any of the explanatory features in the image and essentially defaults to randomly predicting either "medium-high" or "medium-low", resulting in an

AUROC near 0.5; random performance. When using the inverse of the tumor lymphocyte explanation, predictions follow the explanation, leading to worse-than-random performance as shown by AUROC being markedly lower than 0.5. On both X and X_1 the performance is the inverse of the tumor lymphocyte explanation ($0.38 = 1 - 0.62$, $0.31 \approx 1 - 0.67$), while on X_3 this explanation is highly (inversely) predictive with an AUROC of 0.11. Since the AUROC over data subsets follows a similar (but inversed) performance pattern as the normal tumor lymphocyte score, the VLM appears to correctly follow the explanation, independent of how "sensible" it is.

Assumed Predictive Explanations: The two explanations conjectured to be good predictors stand out compared to those designed to be poor predictors, as seen by the considerably higher AUROC, which increases as we test it on only samples with very high and low predictions by the MIL approaching an AUROC of 0.8. The performance is unexpectedly comparable, though. We hypothesize that the coarse explanation is sufficient for our current set-up with coarse binning into "high risk" and "low risk". Figure 2, rightmost panel, shows that the MIL risk logit increases from 6 to 11 as tumor cells intermingle with adipocytes, which both explanations would predict to be "very high" due to the presence of tumor cells.

To conclude, we provide evidence that the evaluation framework is reasonably able to quantitatively distinguish explanations that are conjectured to be better or worse predictors. The VLM generally follows the explanation as shown by the inverse explanations. Questions about the cause of the similarity between the simple tumor lymphocyte explanation and the detailed explanation remain. This enables researchers to test the improvement of a refined explanation made with, for example, the sliding window experiments presented in Sect. 4.

6   Discussion

A lack of good explanations for DL models in medical image analysis may lead to disastrous results. Although the field of XAI is growing, explanations are rarely tested, criticized, and iterated upon. We argue this is in part due to the absence of ideas and methods for the falsification and quantification of explanations. This limits the individual researcher to confidently provide good explanations to discover and prevent unwanted actions by the model.

In this work, we propose that providing a good explanation is necessary and feasible, and take a step towards providing initial conceptual and technical tools for explanation falsification and quantification. We describe what an explanation is and what makes a good explanation. Additionally, we propose that intuitive sliding-window experiments with an AI-integrated image viewer can test components of an explanation. Finally, using a VLM we can quantitatively compare competing explanations, although it may not yet be precise enough to identify the most predictive among a set of multiple predictive explanations.

There are some limitations, however. GPT-4o is not perfect for medical image analysis [5,25], hence poor predictiveness of an explanation may be due to a limitation of the VLM. Increasingly performant reasoning models will likely bridge this gap soon, though. Second, we only test one model on a single small dataset for which we already had a good explanation from other analyses. Extending this to a larget dataset is straightforward, though, since we require patches without labels. Additionally, a good exercise is to apply the sliding window approach to a model while being blind to its task to provide an initial explanation from scratch, and continuously refine it with feedback from the quantitative evaluation, to finally validate the explanation on an external dataset.

In the future, we foresee explanations to be generated and evaluated by a multi-agent system, as shown to work for statistical hypothesis testing and WSI classification [9,11]. Using the concepts presented in this work, we imagine an agent that selects patches that challenges the existing explanation, and another agent to refute incorrect components, iterating until a stable and predictive explanation is found.

With testable explanations that continue to improve, AI in the medical domain can be better understood, reducing the likelihood of unwanted behavior in the clinic. As these models solve harder tasks using multimodal data, we believe good explanations may, in the future, provide novel biological hypotheses about, e.g., prognostic or predictive features.

References

1. Arjovsky, M., Bottou, L., Gulrajani, I., Lopez-Paz, D.: Invariant risk minimization. arXiv preprint arXiv:1907.02893 (2019)
2. Balkenende, L., Teuwen, J., Mann, R.M.: Application of deep learning in breast cancer imaging. Semin. Nucl. Med. **52**(5), 584–596 (2022)
3. Borys, K., et al.: Explainable ai in medical imaging: an overview for clinical practitioners-saliency-based xai approaches. Eur. J. Radiol. **162**, 110787 (2023)
4. Brunekreef, J.: Letter to the editor regarding article "prior to initiation of chemotherapy, can we predict breast tumor response? deep learning convolutional neural networks approach using a breast mri tumor dataset." J. Imaging Inf. Med. **37**(5), 2706–2708 (2024)
5. Cai, X., Zhan, L., Lin, Y.: Assessing the accuracy and clinical utility of gpt-4o in abnormal blood cell morphology recognition. Digit. Health **10**, 20552076241298504 (2024)
6. Chen, R.J., et al.: Pan-cancer integrative histology-genomic analysis via multimodal deep learning. Cancer Cell **40**(8), 865–878 (2022)
7. DeGrave, A.J., Janizek, J.D., Lee, S.I.: Ai for radiographic covid-19 detection selects shortcuts over signal. Nat. Mach. Intell. **3**(7), 610–619 (2021)
8. Geirhos, R., et al.: Shortcut learning in deep neural networks. Nat. Mach. Intell. **2**(11), 665–673 (2020)
9. Ghezloo, F., et al.: Pathfinder: a multi-modal multi-agent system for medical diagnostic decision-making applied to histopathology. arXiv preprint arXiv:2502.08916 (2025)
10. Howard, F.M., Kather, J.N., Pearson, A.T.: Multimodal deep learning: an improvement in prognostication or a reflection of batch effect? Cancer Cell **41**(1), 5–6 (2023)

11. Huang, K., Jin, Y., Li, R., Li, M.Y., Candès, E., Leskovec, J.: Automated hypothesis validation with agentic sequential falsifications. arXiv preprint arXiv:2502.09858 (2025)
12. Ilse, M., Tomczak, J., Welling, M.: Attention-based deep multiple instance learning. In: International Conference on Machine Learning, pp. 2127–2136. PMLR (2018)
13. Jiang, X., et al.: End-to-end prognostication in colorectal cancer by deep learning: a retrospective, multicentre study. Lancet Digit. Health **6**(1), e33–e43 (2024)
14. de Jong, E.D., Marcus, E., Teuwen, J.: Current pathology foundation models are unrobust to medical center differences. arXiv preprint arXiv:2501.18055 (2025)
15. Macenko, M., et al.: A method for normalizing histology slides for quantitative analysis. In: 2009 IEEE International Symposium on Biomedical Imaging: From Nano to Macro, pp. 1107–1110. IEEE (2009)
16. Marcus, E., Teuwen, J.: Artificial intelligence and explanation: How, why, and when to explain black boxes. Eur. J. Radiol. 111393 (2024)
17. McGenity, C., et al.: Artificial intelligence in digital pathology: a systematic review and meta-analysis of diagnostic test accuracy. npj Digit. Med. **7**(1), 114 (2024)
18. Reitsam, N.G., Grosser, B., Steiner, D.F., Grozdanov, V., Wulczyn, E., L'Imperio, V., Plass, M., Müller, H., Zatloukal, K., Muti, H.S., et al.: Converging deep learning and human-observed tumor-adipocyte interaction as a biomarker in colorectal cancer. Commun. Med. **4**(1), 163 (2024)
19. Saha, A., et al.: A machine learning approach to radiogenomics of breast cancer: a study of 922 subjects and 529 dce-mri features. Br. J. Cancer **119**(4), 508–516 (2018)
20. Schirris, Y., Gavves, E., Nederlof, I., Horlings, H.M., Teuwen, J.: Deepsmile: contrastive self-supervised pre-training benefits msi and hrd classification directly from h&e whole-slide images in colorectal and breast cancer. Med. Image Anal. **79**, 102464 (2022)
21. Selvaraju, R.R., Cogswell, M., Das, A., Vedantam, R., Parikh, D., Batra, D.: Gradcam: Visual explanations from deep networks via gradient-based localization. In: Proceedings of the IEEE International Conference on Computer Vision, pp. 618–626 (2017)
22. Van der Velden, B.H., Kuijf, H.J., Gilhuijs, K.G., Viergever, M.A.: Explainable artificial intelligence (xai) in deep learning-based medical image analysis. Med. Image Anal. **79**, 102470 (2022)
23. Wang, X., et al.: Retccl: clustering-guided contrastive learning for whole-slide image retrieval. Med. Image Anal. **83**, 102645 (2023)
24. Wulczyn, E., et al.: Interpretable survival prediction for colorectal cancer using deep learning. NPJ Digit. Med. **4**(1), 71 (2021)
25. Zhang, J., et al.: A comparative study of gpt-4o and human ophthalmologists in glaucoma diagnosis. Sci. Rep. **14**(1), 1–7 (2024)

MATEX: Multi-scale Attention and Text-Guided Explainability of Medical Vision-Language Models

Muhammad Imran[1], Chi Lee[2], and Yugyung Lee[1(✉)]

[1] Computer Science, School of Science and Engineering, University of Missouri, Kansas City, USA
`mi3dr@umkc.edu`, `leeyu@umkc.edu`
[2] Division of Pharmacology and Pharmaceutical Sciences, School of Pharmacy, University of Missouri, Kansas City, USA
`leech@umkc.edu`

Abstract. We introduce MATEX (Multi-scale Attention and Text-guided Explainability), a novel framework that advances interpretability in medical vision-language models by incorporating anatomically informed spatial reasoning. MATEX synergistically combines multi-layer attention rollout, text-guided spatial priors, and layer consistency analysis to produce precise, stable, and clinically meaningful gradient attribution maps. By addressing key limitations of prior methods—such as spatial imprecision, lack of anatomical grounding, and limited attention granularity—MATEX enables more faithful and interpretable model explanations. Evaluated on the MS-CXR dataset, MATEX outperforms the state-of-the-art M2IB approach in both spatial precision and alignment with expert-annotated findings. These results highlight MATEX's potential to enhance trust and transparency in radiological AI applications.

Keywords: Explainable AI · Medical Imaging · Vision-Language Models · Gradient Attribution · Attention Rollout · Chest X-ray · CLIP

1 Introduction

Recent advances in Vision-Language Models (VLMs) such as CLIP [9], BioViL [2], and GLoRIA [6] have demonstrated impressive capabilities in medical imaging tasks, including chest X-ray classification [2]. By bridging visual features with clinical text, these models offer promise for automated diagnosis, triage, and decision support. However, their growing utility is hindered by a fundamental limitation: a lack of interpretability. Without transparent, trustworthy explanations, these models remain ill-suited for high-stakes clinical environments, where accountability, safety, and human oversight are paramount.

© The Author(s), under exclusive license to Springer Nature Switzerland AG 2026
M. Reyes et al. (Eds.): iMIMIC 2025, LNCS 16464, pp. 137–148, 2026.
https://doi.org/10.1007/978-3-032-17611-0_14

In clinical settings, interpretability is not a luxury—it is a prerequisite for adoption. Physicians must be able to trace AI predictions back to meaningful visual evidence, assess whether those predictions align with known pathological findings, and understand when and why models may fail. Furthermore, developers rely on interpretability to uncover latent biases, assess failure modes, and refine model behavior [16]. The ability to generate attribution maps that are both spatially precise and clinically grounded is essential for building trust in AI-assisted diagnostics.

Conventional gradient-based interpretability techniques—such as Saliency Maps [13], Grad-CAM [11], and Integrated Gradients [14]—offer limited insight into model behavior. These methods often suffer from low spatial resolution, high noise sensitivity, and lack of anatomical correspondence. Transformer-based VLMs, while powerful, introduce additional complexity through deep self-attention layers, making it difficult to trace meaningful signal propagation [1,3].

To overcome these challenges, recent approaches such as RISE [8] and M2IB [15] incorporate perturbation-based analysis and information bottlenecks to enhance attribution quality. Despite their progress, these methods still fall short in key areas: they lack anatomical priors, struggle with layer-wise attention inconsistency, and rarely incorporate language-grounded spatial guidance— an essential element for multimodal medical understanding.

We propose **MATEX** (Multi-scale Attention and Text-guided Explainability), a novel framework designed to bridge this gap. MATEX generates anatomically faithful and text-aware attribution maps by leveraging three core innovations: (1) a multi-scale attention rollout strategy to capture signal flow across transformer layers, (2) clinically informed spatial priors derived from radiological text, and (3) a layer consistency mechanism to filter out unstable gradients.

Our contributions are threefold:

- **Multi-Scale Attention Aggregation:** We extend attention rollout to fuse spatial features across transformer layers, improving attribution granularity and robustness.
- **Text-Guided Spatial Priors:** We incorporate structured clinical language to anchor gradient attribution maps around anatomically relevant regions.
- **Layer Consistency Filtering:** We propose a cross-layer consistency scheme to suppress noisy or unstable signals, leading to more reliable and interpretable visual explanations.

Extensive evaluations on the MS-CXR benchmark [2] show that MATEX surpasses existing interpretability methods—including M2IB [15] and Chefer et al. [3]—in both quantitative metrics (confidence change, ROAR+) and qualitative alignment with expert annotations. Our results highlight the importance of integrating textual and architectural priors for clinically trustworthy AI explanations.

2   Related Work

Medical Vision-Language Models. Vision-Language Models (VLMs) are increasingly used in medical imaging to align visual features with clinical text. CLIP [9] catalyzed this trend, inspiring domain-specific adaptations. BioViL [2] fine-tuned vision-language alignment on chest X-rays, while GLo-RIA [6] enhanced spatial grounding of radiological phrases, and Zhou et al. [18] proposed attentive semantic consistency for report generation. Yet, most models remain black boxes, offering limited insight into how clinical descriptions guide visual predictions, thus hindering clinical trust and adoption.

Gradient-Based Attribution Methods. Classic attribution methods like Saliency Maps [13], Grad-CAM [11], and Integrated Gradients [14] highlight pixel importance but often suffer from noise and poor localization. Extensions like CLIP-IG [17] adapt these for multimodal tasks but still lack anatomical grounding. Most approaches also neglect clinical context—e.g., anatomical regions or directional cues—which are critical in radiology.

Attention-Based Interpretability. Attention-based methods visualize internal transformer dynamics. Abnar and Zuidema [1] introduced attention rollout, while Chefer et al. [3] extended attribution to encoder-decoder models. Though model-intrinsic, such maps are often diffuse and not clinically meaningful. M2IB [15] addressed this with information bottlenecks, yet without text or anatomical priors, its interpretability remains limited.

Interpretability in Medical Imaging. Interpretability in medical imaging demands anatomical accuracy and alignment with clinical language. Visual bottlenecks [16] and contrastive methods [5] have improved saliency, but often ignore linguistic cues like "right upper lobe." Moreover, most evaluations remain qualitative, as noted in iMIMIC's call for structured, domain-grounded interpretability benchmarks.

These gaps across model design, attribution, and evaluation motivate our proposed **MATEX** framework, which integrates gradient flow, attention rollout, and text-informed spatial priors to generate anatomically and semantically faithful explanations.

3   Methods

3.1   MATEX Framework Overview

MATEX (Multi-Scale Attention and Text-guided Explainability) is an interpretable framework for medical vision-language models (VLMs), developed to generate clinically grounded and anatomically faithful explanations. The system combines hierarchical transformer attention, gradient attribution, and text-guided spatial priors, as illustrated in Fig. 1.

The architecture consists of five interconnected components:

- **Visual Attention Maps (vmaps):** Derived from the image encoder using value weights, raw attention, and gradient backpropagation. Multi-scale rollout is enhanced via the VR_I module, producing anatomically relevant saliency heatmaps.

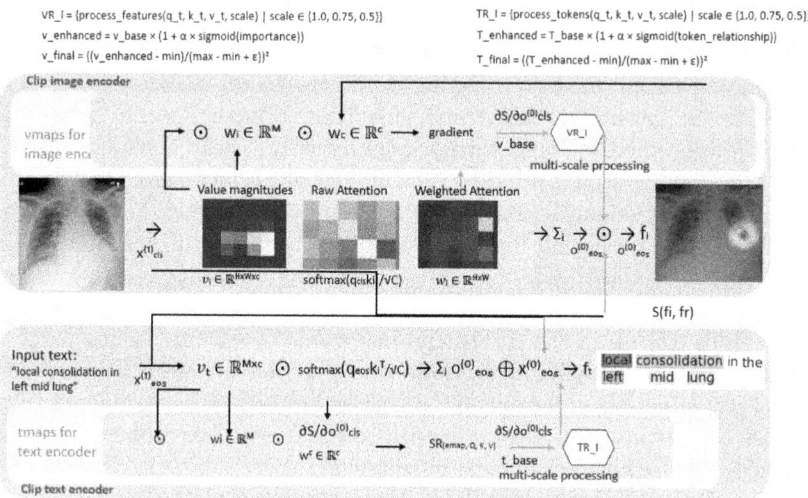

Fig. 1. MATEX Architecture. The image encoder (top) processes a chest X-ray to extract hierarchical attention, value matrices, and gradients for visual attribution. The text encoder (bottom) processes clinical phrases (e.g., "left mid lung") to generate token-level attention and anatomical priors. Visual relevance (VR_I) and text relevance (TR_I) modules perform multi-scale processing. All signals are fused into a final explanation map, guided by attention consistency, gradient sensitivity, and anatomical localization.

- **Textual Attention Maps (tmaps):** From the text encoder, token embeddings are weighted via attention mechanisms and processed by the TR_I module to emphasize clinically important phrases. Token relevance scores are aligned with regions in the image.
- **Gradient Attribution:** Gradients of the image-text similarity score $S(f_i, f_r)$ are computed with respect to both the image features $o_{\text{eos}}^{(0)}$ and the text features $o_{\text{cls}}^{(0)}$. These gradients highlight the sensitivity of specific patches and tokens to the alignment decision.
- **Text-Guided Spatial Priors:** Anatomical references in text (e.g., "left lower lung") are parsed and converted into 2D spatial priors using Gaussian region maps. These soft priors guide visual explanation toward clinically meaningful locations.
- **Multi-Scale Fusion:** All attribution maps—gradient, attention rollout, consistency, and text priors—are combined via a tunable fusion scheme to produce the final explanation map A_{MATEX}.

3.2 Multi-scale Attention Rollout and Consistency

As visualized in Fig. 1, MATEX begins by processing a chest X-ray image through a transformer-based visual encoder (e.g., CLIP ViT). This encoder produces attention maps across multiple layers, capturing hierarchical visual semantics: early layers attend to low-level structural patterns (e.g., lung borders), while deeper layers highlight more abstract clinical features (e.g., lesions or opacities).

To leverage this multi-level information, MATEX performs *multi-scale attention rollout* from the [CLS] token to image patches. This operation traces how attention flows through the network layers, producing a comprehensive attention map A_{flow}:

$$A_{\text{flow}} = \sum_{l=1}^{L} w_l \cdot \text{Attention}^{(l)}_{[\text{CLS}] \to \text{patches}}, \quad w_l = \frac{\exp(\tau l)}{\sum_{k=1}^{L} \exp(\tau k)} \quad (1)$$

Here, L is the total number of transformer layers, and the temperature parameter τ controls the weighting scheme, placing more emphasis on deeper layers when identifying pathology-related semantics.

However, attention scores can vary significantly between layers, particularly in ambiguous or noisy regions. To mitigate this, MATEX introduces a *consistency scoring module* (shown in the figure as a blue consistency block following the attention rollout stream). For each spatial patch i, we compute a reliability score C_i based on the variance of attention values across layers:

$$C_i = \frac{1}{1 + \sigma_l(\text{Attention}^{(l)}_{[\text{CLS}] \to i})} \quad (2)$$

This scoring function penalizes inconsistent patches—those where attention fluctuates across layers—and highlights stable, semantically grounded regions. These scores are later used to modulate the gradient attribution map A_{grad}, as illustrated by the fusion pathway in the figure where $C \odot A_{\text{grad}}$ is computed before entering the fusion module.

Clinical Relevance. For example, in radiographs where both the left and right lungs exhibit subtle opacities, attention across layers may drift. The consistency score C_i suppresses such instability, helping MATEX focus its explanation on robust findings like well-formed consolidations, especially when aligned with gradient or text-derived signals.

By explicitly modeling both the hierarchical attention flow and its cross-layer consistency, this component ensures that explanations are not only attentive to relevant features but also stable across model depths, essential for clinical trustworthiness and reproducibility.

3.3   Text-Guided Spatial Priors

As illustrated in the purple pathway in Fig. 1, MATEX processes the clinical report alongside the image to derive anatomical guidance. The text branch begins with a transformer-based encoder (e.g., CLIP's text encoder), which extracts token-level representations of the report. To transform these linguistic features into spatially relevant guidance, MATEX incorporates a **rule-based parser** that identifies anatomical references, such as "left lower lobe" or "right apex".

These phrases are mapped to a set of anatomical regions R, each associated with a corresponding spatial zone defined by coordinate ranges in normalized (x, y) image coordinates. For example:

- "right apex" $\rightarrow$ upper-right lung region: $x \in [0.5, 1.0], y \in [0.0, 0.4]$
- "left base" $\rightarrow$ lower-left lung region: $x \in [0.0, 0.5], y \in [0.6, 1.0]$

For each region $r \in R$, we define smooth coordinate range weights that create soft transitions within the specified anatomical boundaries. These weights are computed using clamped linear interpolation to ensure smooth falloff at region boundaries:

$$w_x^{(r)} = \mathrm{clamp}\left(\frac{x - x_{\min}}{x_{\max} - x_{\min}}, 0, 1\right) \cdot \mathrm{clamp}\left(\frac{x_{\max} - x}{x_{\max} - x_{\min}}, 0, 1\right) \tag{3}$$

$$w_y^{(r)} = \mathrm{clamp}\left(\frac{y - y_{\min}}{y_{\max} - y_{\min}}, 0, 1\right) \cdot \mathrm{clamp}\left(\frac{y_{\max} - y}{y_{\max} - y_{\min}}, 0, 1\right) \tag{4}$$

The final spatial prior combines contributions from all detected regions and applies max-normalization for stability:

$$M(x, y) = 1 + (\lambda_s - 1) \cdot \mathrm{normalize}\left(\sum_{r \in R} w_x^{(r)} \cdot w_y^{(r)}\right) \tag{5}$$

where $\mathrm{normalize}(\cdot)$ performs min-max normalization: $\mathrm{normalize}(z) = \frac{z - \min(z)}{\max(z) - \min(z)}$ to ensure consistent scaling across different numbers of detected regions.

The parameter $\lambda_s \geq 1$ controls the strength of the text-derived spatial bias: higher values increase the influence of text priors on the final attribution map. When no anatomical text is detected, $M(x, y) = 1$, and the prior becomes neutral.

Integration and Role. As shown in the fusion block of Fig. 1, the spatial prior $M(x, y)$ modulates the gradient map A_{grad} through an elementwise product $M \odot A_{\mathrm{grad}}$, encouraging explanations to focus on medically relevant regions inferred from the text.

Clinical Motivation. For instance, if a radiology report references "opacity in the left lower lung", but the model's raw attribution highlights unrelated areas (e.g., mediastinum), the spatial prior will shift the attribution map toward the correct anatomical zone, improving interpretability and alignment with expert reasoning.

This anatomically aware spatial bias offers domain grounding to the model's explanations and is particularly useful in ambiguous or complex radiographs where gradient-based or attention-based signals alone may lack spatial specificity.

3.4   Multi-component Attribution Fusion

The final output of MATEX is a clinically grounded explanation map A_{MATEX}, computed via a weighted fusion of complementary information sources, as depicted in the fusion block (bottom-center) of Fig. 1. This fusion balances raw gradient relevance, attention-based contextualization, anatomical localization, and layer stability filtering.

Fusion Formula:

$$A_{\mathrm{MATEX}} = \alpha \cdot A_{\mathrm{grad}} + \beta \cdot A_{\mathrm{flow}} + \gamma \cdot (C \odot A_{\mathrm{grad}}) + \delta \cdot (M \odot A_{\mathrm{grad}}) \tag{6}$$

where the weights are configured as:

- $\alpha = 0.5$: Weight for gradient-based saliency, providing fine-grained importance
- $\beta = 0.2$: Weight for multi-scale attention rollout, reflecting semantic attention
- $\gamma = \lambda_c$: Configurable weight for consistency-modulated gradients (typically 0.3–0.4)
- $\delta = 0.2$: Weight for anatomically-guided gradients

The term $(C \odot A_{\mathrm{grad}})$ suppresses unreliable gradient regions with high inter-layer variability. The term $(M \odot A_{\mathrm{grad}})$ reinforces gradient values that overlap with relevant anatomical regions inferred from the report.

Weight Configuration: Unlike traditional normalized weighting schemes, MATEX uses a flexible weighting approach:

$$\alpha + \beta + \delta = 0.9 \quad \text{(fixed components)} \tag{7}$$

$$\gamma = \lambda_c \quad \text{(configurable consistency weight)} \tag{8}$$

This design maintains stable performance for base components while allowing practitioners to adjust the emphasis on layer consistency (γ) based on the reliability requirements of the specific clinical task, without requiring rebalancing of all other components.

Interpretability Role. This fusion enables *clinical flexibility and control*—the user (radiologist or model developer) can emphasize the modality that aligns best with the diagnostic objective. For example: - Emphasizing A_{flow} (increasing β) enhances interpretability through transformer semantics. - Increasing δ favors anatomically-aligned explanations. - Boosting γ enhances attribution robustness by filtering noisy gradients.

4   Experiments and Results

4.1   Dataset and Implementation Details

We evaluate *MATEX* on the MS-CXR benchmark [2], a large-scale chest X-ray dataset featuring radiographic images paired with structured clinical reports containing spatially grounded annotations. The dataset includes localized pathologies such as consolidation, opacity, and pleural effusion, making it ideal for assessing spatial alignment and interpretability.

We use the pretrained ViT-B/32 CLIP encoder as the vision-language backbone. Gradient-based attribution maps are derived from the last 2âĂŞ3 transformer layers, while multi-scale attention rollout aggregates attention across all layers to enhance spatial coverage and semantic richness. Spatial priors are computed from anatomical phrases parsed from the reports, mapped to soft spatial masks. The spatial prior weight λ_s and consistency weight λ_c are tuned on a validation set within the ranges $\lambda_s \in [1.5, 3.5]$ and $\lambda_c \in [0.2, 0.6]$. All experiments are implemented in PyTorch and run on NVIDIA A100 GPUs.

4.2 Evaluation Metrics

To evaluate the effectiveness of *MATEX*, we perform both quantitative and qualitative analyses. **Quantitative Metrics.**

- *Confidence Drop* ($\downarrow$) Measures the reduction in model confidence after masking highly attributed pixels. Lower values suggest the highlighted regions are essential to the model's decision.
- *Confidence Increase* ($\uparrow$) Captures the rise in model confidence when minimally attributed pixels are removed. Higher values indicate more effective identification of irrelevant features.
- *ROAR+* ($\uparrow$) Evaluates attribution quality by retraining the model after masking top-attributed regions. Higher scores imply stronger alignment between predictions and informative input areas.
- *Localization Accuracy* ($\uparrow$) Quantifies the spatial overlap between attribution maps and ground-truth pathology regions. Higher values reflect improved anatomical fidelity and alignment with clinical annotations.

Qualitative Assessment. In addition to quantitative metrics, we examine representative attribution maps to evaluate interpretability and clinical plausibility. This includes assessing whether highlighted regions correspond to anatomically meaningful structures and align with textual descriptors from associated radiology reports (e.g., "right upper lobe nodule"). Visual clarity, spatial precision, and consistency with expected pathology patterns are reviewed to determine how well attribution outputs support clinician reasoning and explainability in real-world diagnostic settings.

4.3 Quantitative Results

Tables 1 and 2 present quantitative evaluations on the MS-CXR dataset using both image-based and text-based metrics. We report three commonly used attribution quality indicators: confidence drop (lower is better), confidence increase (higher is better), and ROAR+ (higher is better), assessing both image and textual modalities.

Findings. MATEX achieves the best confidence increase (47.55%) and lowest drop (0.51%) in the image modality, indicating robust, decision-aligned attributions that surpass all baselines, including M2IB. For image ROAR+, MATEX ranks second (37.9%) just below M2IB (38.7%) but well ahead of methods like Saliency (25.46%) and Chefer (24.42%). In text-based evaluations, MATEX maintains competitive performance, with strong ROAR+ (16.98%), low drop (1.91%), and high gain (50.07%). Although RISE achieves slightly better confidence increase (57.2%) and lower drop (1.16%), MATEX's higher ROAR+ (16.98% vs 12.09%) indicates more reliable attributions. Overall, MATEX consistently ranks at the top across both modalities, demonstrating its ability to deliver coherent, reliable explanations in vision-language medical models.

Table 1. Quantitative evaluation on MS-CXR âĂŞ **Image metrics**. Best per column in bold.

Method	Conf. Drop ↓	Conf. Incr. ↑	ROAR+ ↑
GradCAM [10]	2.76 ± 0.03	12.64 ± 0.46	3.54 ± 0.80
Saliency [12]	0.81 ± 0.01	35.08 ± 0.44	25.46 ± 1.35
KS [7]	2.37 ± 0.04	10.24 ± 0.68	12.67 ± 1.02
RISE [8]	3.94 ± 0.03	7.28 ± 0.44	16.79 ± 0.76
Chefer et al. [4]	1.87 ± 0.02	21.44 ± 0.46	24.42 ± 1.19
M2IB [15]	0.55 ± 0.01	45.92 ± 0.70	**38.7 ± 0.86**
MATEX (Ours)	**0.51 ± 0.03**	**47.55 ± 1.10**	37.9 ± 0.79

Table 2. MS-CXR Text Metrics. Best results per column in bold.

Method	Conf. Drop ↓	Conf. Incr. ↑	ROAR+ ↑
GradCAM [10]	2.26 ± 0.04	36.24 ± 0.54	11.07 ± 0.62
Saliency [12]	3.35 ± 0.03	18.88 ± 0.54	15.79 ± 0.92
KS [7]	2.40 ± 0.05	34.12 ± 0.77	14.28 ± 1.09
RISE [8]	**1.16 ± 0.02**	**57.2 ± 0.65**	12.09 ± 1.52
Chefer et al. [4]	2.93 ± 0.03	28.08 ± 0.34	9.11 ± 0.60
M2IB [15]	2.28 ± 0.04	35.48 ± 0.69	16.31 ± 0.75
MATEX (Ours)	1.91 ± 0.06	50.07 ± 0.46	**16.98 ± 0.66**

4.4 Qualitative Evaluation

Figure 2 shows a qualitative comparison between *MATEX* and *M2IB* on the MS-CXR dataset. Each column depicts a clinical case, with *MATEX* heatmaps on the top row and *M2IB* results below, visualized as overlayed attribution maps on chest X-rays.

Findings. Figure 2 illustrates that MATEX offers several consistent improvements over M2IB. Across all evaluated cases, MATEX produces heatmaps with sharper localization and tighter alignment to pathological regions, enhancing anatomical coherence. In scenarios involving bilateral or multi-zone findings— such as lower lobe consolidation—MATEX accurately captures both affected zones while avoiding overspill, whereas M2IB often yields diffuse attention. Additionally, MATEX exhibits improved boundary adherence, particularly in cases of opacification, which enhances interpretability for clinical use. It also demonstrates reduced noise activation by limiting attribution to relevant structures, unlike M2IB, which sometimes highlights unrelated areas.

These improvements are attributable to MATEX's ability to produce more anatomically precise and clinically aligned explanations. The framework effectively captures multi-region pathologies without spatial drift and maintains clear boundaries around relevant findings.

5   Discussion and Limitations

Performance Summary. *MATEX* improves interpretability in medical vision-language models by combining multi-scale gradient fusion, anatomical priors, and text-guided spatial consistency. Quantitative (Tables 1, 2) and qualitative (Fig. 2) results show that MATEX outperforms prior methods, achieving higher confidence increase and ROAR+ with minimal confidence drop. It also yields

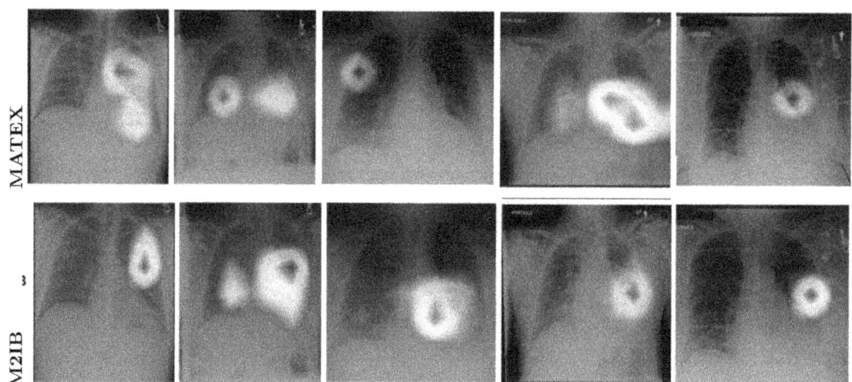

Fig. 2. Qualitative comparison between MATEX and M2IB on MS-CXR. Each column shows a clinical case with **MATEX (Top)** and **M2IB (Bottom)**. **Left to Right:** (1) *Left upper lobe consolidation*: MATEX localizes sharply; M2IB is diffuse. (2) *Bilateral lower lobe consolidation*: MATEX captures both lobes; M2IB is less focused. (3) *Right mid-to-upper zone opacity*: MATEX aligns closely with pathology. (4) *Left lower lobe opacification*: MATEX offers better boundary definition. (5) *Left mid-lung consolidation*: MATEX maintains anatomical fidelity. Overall, MATEX shows superior spatial precision and clinical alignment.

finer-grained localization and stronger anatomical alignment than M2IB, supporting its clinical relevance where spatial precision and trust are vital.

Key Strengths. MATEX enables anatomically grounded attributions through spatial priors, captures both low- and high-level features via multi-scale attention, and fuses gradients with attention for rich explanations. Cross-modal fidelity keeps visual outputs aligned with clinical semantics.

Limitations. MATEX adds $\sim$40% overhead from multi-layer backpropagation. Its priors are currently specific to chest X-rays and require adaptation for other modalities. Attribution quality is sensitive to parameter tuning and the reliability of CLIP attention.

6   Conclusion

We introduced *MATEX*, a framework that enhances interpretability in medical vision-language models by integrating multi-scale attention, anatomically-informed spatial priors, and consistency-driven refinement. By combining gradient signals with layer-wise attention rollout and clinical spatial cues, MATEX produces high-fidelity, anatomically grounded explanations for CLIP-based models. Experiments on the MS-CXR dataset show that MATEX outperforms Grad-CAM, RISE, and M2IB across both image- and text-based attribution metrics, with qualitative results confirming improved spatial precision and diagnostic relevance. These results demonstrate that incorporating textual semantics and spatial priors not only improves explanation quality but also aligns model outputs with clinical reasoning. While maintaining computational efficiency, MATEX addresses limitations in gradient-based methods such as spatial incoherence and modality misalignment.

References

1. Abnar, S., Zuidema, W.: Quantifying attention flow in transformers. In: Proceedings of the 58th Annual Meeting of the Association for Computational Linguistics, pp. 4190–4197 (2020)
2. Boecking, B., et al.: Making the most of text semantics to improve biomedical vision-language processing. In: Computer Vision–ECCV 2022: 17th European Conference, pp. 387–405 (2022)
3. Chefer, H., Gur, S., Wolf, L.: Generic attention-model explainability for interpreting bi-modal and encoder-decoder transformers. In: Proceedings of the IEEE/CVF International Conference on Computer Vision, pp. 397–406 (2021)
4. Chefer, H., Gur, S., Wolf, L.: Transformer interpretability beyond attention visualization. In: Proceedings of the IEEE/CVF Conference on Computer Vision and Pattern Recognition, pp. 782–791 (2021)
5. Ma, Y., Bertsimas, D.: M3H: multimodal multitask machine learning for healthcare. In: Advancements in Medical Foundation Models Workshop, NeurIPS (2024)
6. Huang, S.-C., Shen, L., Lungren, M.P., Yeung, S.: Gloria: a multimodal global-local representation learning framework for label-efficient medical image recognition. In: Proceedings of ICCV, pp. 3942–3951 (2021)

7. Lundberg, S.M., Lee, S.-I.: A unified approach to interpreting model predictions. In: Advances in Neural Information Processing Systems, pp. 4765–4774 (2017)
8. Petsiuk, V., Das, A., Saenko, K.: Rise: randomized input sampling for explanation of black-box models. In: British Machine Vision Conference (2018)
9. Radford, A., et al.: Learning transferable visual models from natural language supervision. In: Intern. Conference on Machine Learning, pp. 8748–8763 (2021)
10. Selvaraju, R.R., Cogswell, M., Das, A., Vedantam, R., Parikh, D., Batra, D.: Grad-cam: visual explanations from deep networks via gradient-based localization. In: Proceedings of the IEEE International Conference on Computer Vision, pp. 618–626 (2017)
11. Selvaraju, R.R., Cogswell, M., Das, A., Vedantam, R., Parikh, D., Batra, D., et al.: Grad-cam: visual explanations from deep networks via gradient-based localization. Int. J. Comput. Vis. **128**(2), 336–359 (2020)
12. Simonyan, K., Vedaldi, A., Zisserman, A.: Deep inside convolutional networks: visualising image classification models and saliency maps. arXiv preprint arXiv:1312.6034 (2013)
13. Simonyan, K., Zisserman, A.: Deep inside convolutional networks: visualising image classification models and saliency maps. In: International Conference on Learning Representations (2014)
14. Sundararajan, M., Taly, A., Yan, Q.: Axiomatic attribution for deep networks. In: Proceedings of the 34th International Conference on Machine Learning, volume 70 of PMLR, pp. 3319–3328 (2017)
15. Mai, S., Zeng, Y., Hu, H.: Multimodal information bottleneck: learning minimal sufficient unimodal and multimodal representations. IEEE Trans. Multimedia **25**, 4121–4134 (2023)
16. Baniecki, H., Biecek, P., Arnaiz-González, Á., et al.: Towards evaluating explanations of vision transformers for medical imaging. In: MICCAI (2023)
17. Zhao, J., Rodriguez, M., Chen, Y.: Gradient-based visual explanations in medical transformer networks. In: Proceedings of the Conference on Computer Vision and Pattern Recognition, pp. 12745–12754 (2024)
18. Zhou, Y., Huang, L., Zhou, T., Fu, H., Shao, L.: Visual-textual attentive semantic consistency for medical report generation. In: Proceedings of the IEEE/CVF International Conference on Computer Vision, pp. 3985–3994 (2021)

ProtoEFNet: Dynamic Prototype Learning for Inherently Interpretable Ejection Fraction Estimation in Echocardiography

Yeganeh Ghamary[1(✉)], Victoria Wu[2], Hooman Vaseli[2], Christina Luong[3], Teresa Tsang[3], Siavash A. Bigdeli[1], and Purang Abolmaesumi[2]

[1] Department of Applied Mathematics and Computer Science, Technical University of Denmark, Kongens Lyngby, Denmark
`s194258@dtu.dk` , `sarbi@dtu.dk`
[2] Department of Electrical and Computer Engineering, The University of British Columbia, Vancouver, BC, Canada
`victoriawu@ece.ubc.ca, hoomanv@ece.ubc.ca` , `purang@ece.ubc.ca`
[3] Vancouver General Hospital, Vancouver, BC, Canada

Abstract. Ejection fraction (EF) is a crucial metric for assessing cardiac function and diagnosing conditions such as heart failure. Traditionally, EF estimation requires manual tracing and domain expertise, making the process time-consuming and subject to inter-observer variability. Most current deep learning methods for EF prediction are black-box models with limited transparency, which reduces clinical trust. Some post-hoc explainability methods have been proposed to interpret the decision-making process after the prediction is made. However, these explanations do not guide the model's internal reasoning and therefore offer limited reliability in clinical applications. To address this, we introduce ProtoEFNet, a novel video-based prototype-learning model for continuous EF regression. The model learns dynamic spatio-temporal prototypes that capture clinically meaningful cardiac motion patterns. Additionally, the proposed Prototype Angular Separation (PAS) loss enforces discriminative representations across the continuous EF spectrum. Our experiments on the Echonet-Dynamic dataset show that ProtoEFNet can achieve accuracy on par with its non-interpretable counterpart while providing clinically relevant insight. The ablation study shows the proposed loss boosts the performance with a 2% increase in F1 score from 77.67 ± 2.68 to 79.64 ± 2.10. Our source code is available at: https://github.com/DeepRCL/ProtoEF.

Keywords: Ultrasound · Echocardiography · Ejection Fraction · Regression · Explainable AI · Prototypical Neural Networks

Y. Ghamary and V. Wu are joint first authors. T. Tsang, S. A. Bigdeli and P. Abolmaesumi are joint senior authors. Work done during Y. Ghamary's external stay at UBC.

© The Author(s), under exclusive license to Springer Nature Switzerland AG 2026
M. Reyes et al. (Eds.): iMIMIC 2025, LNCS 16464, pp. 149–159, 2026.
https://doi.org/10.1007/978-3-032-17611-0_15

1   Introduction

Heart failure is a major global health issue that requires accurate and timely assessment for effective management. A key measure of cardiac function is ejection fraction (EF), which quantifies the percentage of blood pumped from the left ventricle with each heartbeat. EF plays a central role in diagnosing heart failure, guiding treatment, and predicting outcomes [7,13]. It is typically measured using echocardiography, where clinicians manually trace the left ventricle at two key points in the cardiac cycle: end-diastole (ED) and end-systole (ES), to estimate volume changes [2]. However, this process is highly operator-dependent, with inter-observer variation ranging from 7.6 to 13.9%, and requires significant expertise [19]. Automating ejection fraction estimation through artificial intelligence can enhance consistency, reduce clinician workload, and support large-scale screening.

Despite progress in automated EF estimation [9,12,14–17,19,20], current methods face key challenges. Many rely on black-box deep learning models, offering limited transparency and reducing clinical trust. Explainability in medical AI is crucial, as clinicians require interpretable reasoning behind model predictions to facilitate adoption in medical settings. However, the existing explainable approaches use post-hoc techniques, such as attention weights [1,16], or gradients [21,23], that interpret the decision-making process after the prediction is made. These explanations are often inconsistent and do not inform the model's internal reasoning, limiting their reliability in clinical applications.

To address these challenges, ante-hoc XAI methods have been introduced, embedding explainability directly into model architectures. Prototype-based models [3,10,11,25] are key examples, where networks learn class-specific prototypes that capture discriminative patterns explicitly used for classification. Unlike post-hoc methods that rely on abstract signals such as attention weights or gradients, which can be diffused, multi-layered, and difficult to interpret [3,22], prototypes offer more explicit, case-based explanations by grounding decisions in example-like visual features, mimicking a clinician's decision process. However, adapting prototype learning to ejection fraction estimation poses specific challenges. EF is a continuous value, requiring a regression-based approach, and unlike other assessments that use static images, echocardiographic assessment depends on capturing spatio-temporal information across frames.

Some prior work has extended prototype-based models to regression [5,6] and video classification [25]. InsightRNet [6] applies prototype learning to regression, but uses discrete ordinal labels rather than truly continuous targets. Furthermore, [5,6] rely on fixed-size patch-based prototypes, which are insufficient for capturing the complex and spatially diverse visual features of ejection fraction (EF), particularly the broad and varying motion of the LV wall.

We introduce ProtoEFNet (Fig. 1), the first video-based prototype model for continuous regression and the first inherently interpretable approach to EF estimation. Our key contributions are: (1) learning dynamic spatio-temporal prototypes that capture clinically relevant motion patterns; (2) proposing a prototype angular separation (PAS) loss to enforce discriminative representations across

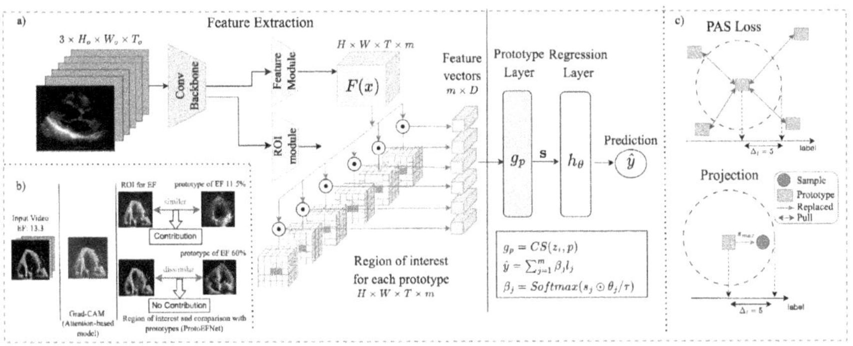

Fig. 1. **(a)** An Overview of the architecture of ProtoEFNet. The feature extractor uses an ROI module to focus on clinically relevant spatio-temporal regions, and the final prediction is the weighted sum of the prototype labels, **(b)** Grad-CAM attention of CoReEcho [14], and the decision process of ProtoEFNet. The activation maps on the input data and prototypes show where the model "looks at" when calculating the cosine similarity, **(c)** The Prototype Angular Separation (PAS) loss increases separation between prototypes with different EF ranges, and prototypes are projected to/replaced with the closest training sample within its EF range.

the continuous EF spectrum; (3) achieving state-of-the-art explainability on the EchoNet-Dynamic dataset, and accuracy on par with leading black-box models; (4) demonstrating through visualizations that ProtoEFNet uniquely attends to essential cardiac features—such as LV wall motion and reduced mitral valve movement in EF< 40% cases—whereas leading black-box methods produce diffuse or clinically irrelevant attention.

2   Methods

Problem Definition. We define a fixed number of m learnable prototype vectors $\mathcal{P} = \{p_1, p_2, p_3, ..., p_m\}$, each associated with a continuous EF label $\mathbf{l} = \{l_1, ..., l_m\}$ and an *importance score* $\theta = \{\theta_1, ..., \theta_m\}$ that is learned during training. Based on the *similarity scores* and the learned importance of each prototype, we generate a score sheet with scores indicating *prototype contributions*. Unlike in conventional classification tasks, where multiple prototypes are associated with each discrete class and scores are calculated at the class level [3], our approach assigns probabilities to individual prototypes, capturing the continuous nature of the label space.

Feature Extractor has a similar architecture to [25], consisting of a pre-trained R(2+1)D-18 backbone, a feature module $F(.)$, and a Region of Interest (ROI) module $M_{p_k}(.)$. Given an input video $x \in \mathbb{R}^{H_o \times W_o \times T_o \times 3}$ with T_o frames, the learned features are $F(x) \in \mathbb{R}^{H \times W \times T \times D}$, and the ROI module generates P occurrence maps $M_{p_k}(x) \in \mathbb{R}^{H \times W \times T}$ highlighting which regions of $F(x)$ are relevant when compared to p_k. Occurrence maps highlight the regions in space and time where it is likely to observe relevant features for EF prediction. We perform a weighted average pooling of the spatiotemporal features using the learned occurrence maps as weights.

Prototype Layer. We use cosine similarity (CS) score as a similarity function between the features $f_{p_k}(x)$ and prototypes p_k, which are both D-dimensional vectors $s_k = g_{p_k}(f_{p_k}(x)) = CS(f_{p_k}(x), p_k)$.

Regression Layer is a linear layer with m weights denoted by θ. Unlike class-based prototype models, the final prediction is calculated as a weighted average of the prototype labels:

$$\hat{y} = \sum_{k=1}^{m} \beta_k \cdot l_k, \quad \beta_k(x) = \frac{e^{(s_k * \theta_k)/\tau}}{\sum_{i=1}^{m} e^{(s_k * \theta_k)/\tau}}, \tag{1}$$

where $\beta_k(.)$ indicates the *contribution* of each prototype p_k in the final prediction and is calculated using the softmax function of scaled similarity scores. To encourage *sparsity* explanations, we use a small temperature parameter of $\tau = 0.2$. This ensures that dissimilar prototypes have 0 contribution to the final prediction (see Fig. 1(b)). The overview of ProtoEFNet can be seen in Fig. 1(a).

Training Algorithm. The feature extractor, ROI module, regression layer, and prototype vectors are trained jointly. In the last epoch, the prototypes (and their labels) are *projected* (replaced) onto the closest training feature (see Fig. 1(c)). This allows us to visualize the prototypes, enhancing human-level explainability. We adapt the notion of class-based projection to the regression task using a threshold of Δ_l. It is mathematically defined as:

$$p_j \leftarrow argmax_z CS(z, p_j), \quad z = f_{p_j}(x_i), |y_i - l_j| < \Delta_l. \tag{2}$$

Latent Space Regularization. To learn representative prototypes, the latent space is learned using two distance-based losses. A regression-based cluster loss and prototype sample distance loss (PSD) [6]. Cluster loss creates clusters around prototypes by pulling the samples with similar EF labels using a threshold of Δ_l, and PSD loss ensures that there is at least one sample close to each prototype:

$$\mathcal{L}_{Clst} = -\frac{1}{n} \sum_{i=1}^{n} \max_{p \in \mathcal{P}^c}(s_p(i)), \quad |y_i - l_j| < \Delta_l, \tag{3}$$

$$\mathcal{L}_{PSD} = -\frac{1}{m} \sum_{j=1}^{m} log(1 - \min_{i \in [1,n]} \frac{d_{i,j}}{d_{max}}), \tag{4}$$

where n is the batch size, kmax is the maximum operation selecting the k closest prototypes to each sample, and $s_p(i)$ is the cosine similarity score between the prototype p and sample i. $\mathcal{P}^c$ visualizes the set of prototypes in the vicinity of the sample in the label space. d_{max} is the maximum possible distance in the latent space, which is 2 for the cosine distance score, and $d_{i,j}$ is the cosine distance between sample i and prototype j.

Prototype models [3, 10] have been effective in classification by using separation losses to enforce clear boundaries between class-specific prototypes. However, in regression tasks, where outputs are continuous and ordered, preserving the ordinality and smooth transitions between clusters is essential. In the meantime, prototype vectors representing different EF regions should be further apart than those prototypes belonging to the same EF range. This ensures ordinality in the embedding space and that prototypes are semantically meaningful. However, in practice, we observed that even the most distant prototypes with EF labels of 15% and 86% have similar embedding vectors with a cosine similarity of around 0.7, suggesting that the embeddings lack meaningful distinction.

To address this, and inspired by [11], we propose a novel prototype angular separation loss (PAS) that promotes greater inter-prototype distinction while respecting the continuity between the cluster of the data points required for regression tasks. The PAS loss addresses this by pulling prototypes from different EF regions further apart (see Fig. 1(c). To measure the distance between prototypes, we utilize angular similarity—a monotonic transformation of cosine similarity that ranges from 0 to 1. The PAS loss can be defined as follows:

$$\mathcal{L}_{PAS} = -\frac{1}{m} \sum_{i=1}^{m} \left[\frac{1}{|\mathcal{P}^{\bar{c}}|} \sum_{j \in \mathcal{P}^{\bar{c}}} \log(1 - AS(p_i, p_j)) \right], \quad \mathcal{P}^{\bar{c}} = \{p_j \mid |l_i - l_j| > \Delta_l\}, \tag{5}$$

where AS indicates angular similarity defined as:

$$AS(\mathbf{p_i}, \mathbf{p_j}) = 1 - \frac{1}{\pi} \arccos(CS(\mathbf{p_i}, \mathbf{p_j})). \tag{6}$$

$\mathcal{P}^{\bar{c}}$ is the set of prototypes that have EF label larger than Δ_l of the given prototype. Inside the bracket is the average of the log of the pairwise angular similarity. We chose the average due to its superiority over the max function.

Occurrence Map Regularizor. To further emphasize disease-specific areas, we incorporate the LV segmentation mask into the L1 regularization, penalizing activations outside the LV and reducing attention to irrelevant regions like the background. The overall cost function is defined below, where λ represents weight of each loss term:

$$\mathcal{L} = \lambda_{MSE}\mathcal{L}_{MSE} + \lambda_{Clst}\mathcal{L}_{Clst} + \lambda_{PSD}\mathcal{L}_{PSD} + \lambda_{PAS}\mathcal{L}_{PAS} + \lambda_{Occur}\mathcal{L}_{Occur}. \tag{7}$$

3 Experiment

Dataset and Implementation. EchoNet-Dynamic [18] is the largest public echocardiogram dataset, containing 10,036 apical four-chamber (AP4) videos (112×112 resolution) with varying lengths, each labeled with a single EF value, LV segmentation, and end-systolic/diastolic frame indices. We follow the train-val-test split from [19] and frame sampling from [16]. We augment the data using random rotation. Label balance was achieved by oversampling the minority region (EF < 50%), as addressing imbalance was beyond the scope of this study. The EF value outside the range of [10%, 90%] is clinically uncommon, and the dataset does not include any samples outside this range. Derived from the hyperparameter selection, each video is sampled as a clip of 64 frames (sampling period of 1) with the initial frame index sampled uniformly.

All experiments were conducted using an NVIDIA A100 (40 GB) with PyTorch 2.0.1 and CUDA 12.8. Following hyperparameter tuning, we used 40 prototypes, a softmax temperature of $\tau = 0.2$ for regression layer, $\Delta_l = 5.0\%$, and $k = 3$ in $\mathcal{L}_{Clst}$. The prototype vectors were initialized randomly. Regression layer weights θ_h were initialized to 1, assigning equal importance to prototypes, and prototype labels were uniformly set from 10% to 90%. This is to ensure that prototypes represent all regions of the label space, including low-density regions. We used Adam optimizer with a learning rate of $1e - 4$ for the backbone and regression layer, $1e - 3$ for feature and ROI modules and $3e - 3$ for prototype vectors. We use pretrained weights of [8] and fine-tuned jointly for 30 epochs with a batch size of 16.

Comparison with SOTA Models. In Table 1, we compare ProtoEFNet with SOTA methods on EchoNet-Dynamic. We excluded [5,6], as adapting them to the dataset required major changes that altered their core architecture. We furthermore report the F1 score for the task of indicating whether EF values are lower than 40%, which is a strong indicator of heart failure [19]. ProtoEFNet outperforms most of the SOTA models, including the post-hoc explainable model [16] and EchoCoTr with same frame size. The performance gap between ProtoEFNet and CoReEcho [14] could potentially be narrowed with further training, and more extensive hyperparameter tuning. A key advantage of ProtoEFNet is its inherent explainability, offering transparent insights into its predictions.

Explainability Analysis. Figure 1(b) illustrates ProtoEFNet's decision process for a video sample, a more detailed description of the activation map visualisation can be found in [10]. ProtoEFNet is *inherently interpretable* and *transparent*, with predictions directly linked to prototype contribution and similarity scores. Its explanations are *sparse* and *faithful*, relying only on prototypes with EF values close to the true label—prototypes with distant EF values (e.g., 60%) do not influence the prediction for the sample with EF of 13%. Activation maps on the input video demonstrate both spatial (anatomical localization) and temporal (motion) explainability, highlighting *clinically relevant* features such as reduced LV wall motion and mitral valve movement during systole. Prototypical features shown as activation maps on top of prototypical cases show distinct EF-related patterns: the 60% EF prototype exhibits healthy LV and MV motion, while the 11% EF prototype shows reduced LV motion and a thin LV wall, both clinically meaningful. In Fig. 2, CoReEcho's spatio-temporal attention (Grad-CAM [21]) is compared to ProtoEFNet's activation map. ProtoEFNet demonstrates superior localization, focusing on key structures like the LV wall and mitral/aortic valves, while CoReEcho and EchoCoTr highlight non-specific or clinically irrelevant regions, such as background (see [14]). ProtoEFNet also learns the periodic nature of echo videos and correctly aligns the spatio-temporal features of the input clip to those of the prototype, seen in Fig. 3.

Table 1. The quantitative results on the EchoNet-Dynamic test set. **P** indicates the frame sampling period. Most approaches average clip-level predictions over the entire video (**EV**), while others use a single heartbeat (**SH**). Results marked with (*) denote reproduced performance, while (^{1}) indicates the SOTA method with the frame sampling similar to ours.

Method	Frames	P	Clips	R2 $\uparrow$	MAE $\downarrow$	RMSE $\downarrow$	$F1_{<40\%}\uparrow$	Explainable
Reynaud et al. [20]	128	1	SH	52	5.59	8.38	NA	✗
Bayesian [9]	32	1	SH	75	4.46	NA	77	✗
EchoCoTr [17][1]	36	2	EV	79	4.18	5.59	NA	✗
EchoCoTr [17]	36	4	EV	81	3.98	5.34	NA	✗
CoReEcho [14]	36	1	3 Clips	**82**	**3.90**	**5.13**	**80**	✗
GEMTrans [15][1]	64	1	EV	79	4.15	NA	NA	✗
Resnet2+1D [19][1]*	64	1	EV	80	4.10	5.47	78	✗
EchoGNN [16][1]	64	1	EV	76	4.45	NA	**78**	post-hoc
ProtoEFNet (ours)	64	1	EV	**80**	**4.07**	**5.47**	78	✓

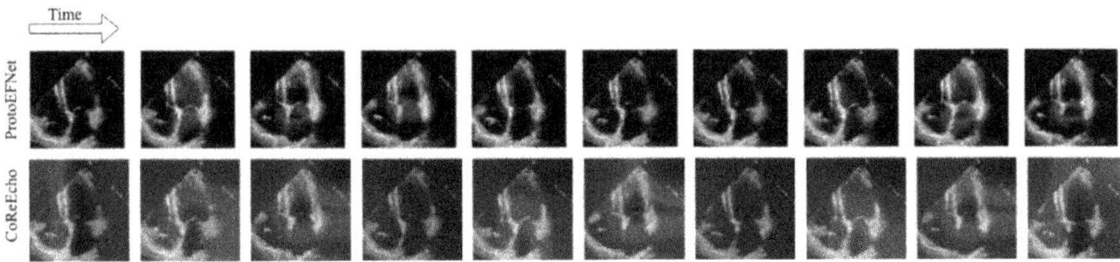

Fig. 2. Grad-CAM [21] of CoReEcho (bottom row) and the activation map of ProtoEFNet (top row). ProtoEFNet is localised on LV wall motion and mitral valve movements during systole (contraction).

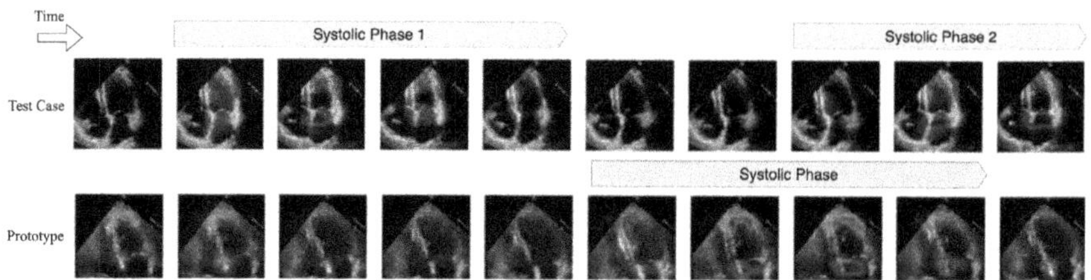

Fig. 3. Activation maps of a test case and the top contributing prototype. The model captures periodic patterns and aligns spatio-temporal features. In this example, it assigns a high similarity score as it "looks at" the LV wall during systole in the input and identifies that it "looks like" the LV wall of the prototype during the same phase.

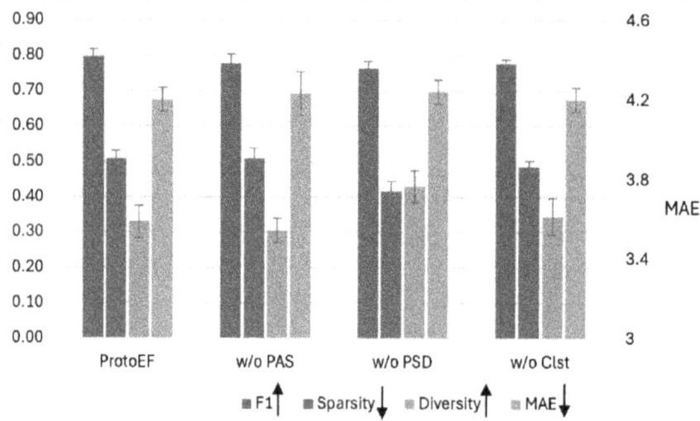

Fig. 4. Ablation Study of different loss components on validation set. Standard deviation is calculated across 5 repetitions of each experiment.

Ablation Study. Figure 4 shows an ablation study of different loss components. Including all loss components yields the best regression performance based on MAE and F1 scores. We evaluate prototype quality using Sparsity and Diversity metrics [6] scaled by the number of prototypes: effective explanations rely on the contribution of a few prototypes (*low Sparsity*), but different predictions rely on different prototypes (*high Diversity*). To demonstrate the effect of these components in the embedding space, we visualize 2D PCA plot of the prototypes and the 100 closest validation features in Fig. 5. Removing $\mathcal{L}_{PSD}$ improves diversity and sparsity but degrades regression performance, as reflected by F1 and MAE scores. Moreover, some prototypes become outliers with no nearby training samples (see Fig. 5). Without $\mathcal{L}_{Clst}$, the samples and prototypes are scattered in the embedding space without any clear ordinality. $\mathcal{L}_{PAS}$ decreases MAE from 4.23 ± 0.11 to 4.20 ± 0.06, and increases F1 score from 77.67 ± 2.68 to 79.64 ± 2.10, while producing more distinct prototypes and ordinally structured clusters in the embedding space.

4   Conclusion

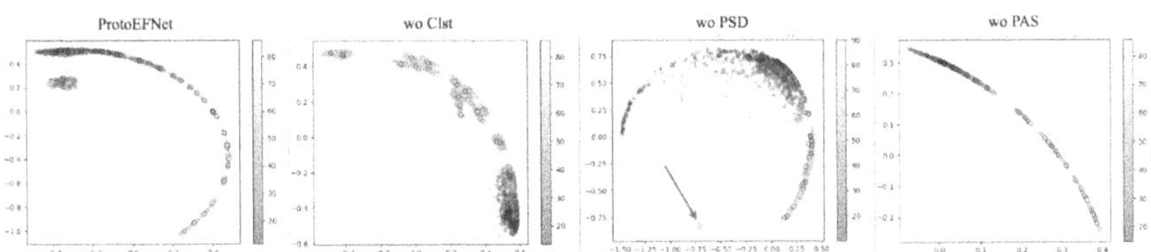

Fig. 5. The PCA plots of the prototypes (circles) and the top 100 closest latent features of the validation set to each prototype (stars). The colors indicate the ground truth EF values.

We proposed ProtoEFNet, the first prototype-based model for video-based continuous EF regression. ProtoEFNet has superior performance to the *post-hoc* explainable model [16] and a performance on par with the black-box models, while offering inherent interpretability, transparency, and clinically meaningful explanations. Qualitative analysis shows superior focus on key cardiac structures, and ablation results confirm the effectiveness of the proposed Prototype Angular Separation loss. Future work will address prototype learning for uncommon EF ranges (minority region).

Acknowledgements. This work was supported in part by the Technical University of Denmark's Travel Grant, Marie og Anders Manssons Memorial Grant, the Canadian Institutes of Health Research (CIHR), and the Natural Sciences and Engineering Research Council of Canada (NSERC). Resources were provided through DTU Computing Center at Technical University of Denmark [4] and Advanced Research Computing at the University of British Columbia [24].

References

1. Bahdanau, D., Cho, K., Bengio, Y.: Neural machine translation by jointly learning to align and translate. arXiv preprint arXiv:1409.0473 (2014)
2. Bamira, D., Picard, M.: Imaging: echocardiology—assessment of cardiac structure and function (2018)
3. Chen, C., Li, O., Tao, D., Barnett, A., Rudin, C., Su, J.K.: This looks like that: deep learning for interpretable image recognition. In: Advances in Neural Information Processing Systems, vol. 32 (2019)
4. DTU Computing Center: DTU computing center resources (2024). https://doi. org/10.48714/DTU.HPC.0001
5. Hesse, L.S., Dinsdale, N.K., Namburete, A.I.: Prototype learning for explainable brain age prediction. In: Proceedings of the IEEE/CVF Winter Conference on Applications of Computer Vision, pp. 7903–7913 (2024)

6. Hesse, L.S., Namburete, A.I.: Insightr-net: interpretable neural network for regression using similarity-based comparisons to prototypical examples. In: International Conference on Medical Image Computing and Computer-Assisted Intervention, pp. 502–511. Springer (2022)
7. Huang, H., et al.: Accuracy of left ventricular ejection fraction by contemporary multiple gated acquisition scanning in patients with cancer: comparison with cardiovascular magnetic resonance. J. Cardiovasc. Magn. Reson. **19**(1), 34 (2016)
8. Kay, W., et al.: The kinetics human action video dataset. arXiv preprint arXiv:1705.06950 (2017)
9. Kazemi Esfeh, M.M., Luong, C., Behnami, D., Tsang, T., Abolmaesumi, P.: A deep bayesian video analysis framework: towards a more robust estimation of ejection fraction. In: International Conference on Medical Image Computing and Computer-Assisted Intervention, pp. 582–590. Springer (2020)
10. Kim, E., Kim, S., Seo, M., Yoon, S.: Xprotonet: diagnosis in chest radiography with global and local explanations. In: Proceedings of the IEEE/CVF Conference on Computer Vision and Pattern Recognition, pp. 15719–15728 (2021)
11. Kraft, S., Broelemann, K., Theissler, A., Kasneci, G., Esslingen am Neckar, G., AG, S.H.: Sparrow: semantically coherent prototypes for image classification. In: Bmvc, p. 186 (2021)
12. Lai, S., et al.: Echomen: combating data imbalance in ejection fraction regression via multi-expert network. In: International Conference on Medical Image Computing and Computer-Assisted Intervention, pp. 624–633. Springer (2024)
13. Loehr, L.R., Rosamond, W.D., Chang, P.P., Folsom, A.R., Chambless, L.E.: Heart failure incidence and survival (from the atherosclerosis risk in communities study). Am. J. Cardiol. **101**(7), 1016–1022 (2008)
14. Maani, F.A., Saeed, N., Matsun, A., Yaqub, M.: Coreecho: continuous representation learning for 2d+ time echocardiography analysis. In: International Conference on Medical Image Computing and Computer-Assisted Intervention, pp. 591–601. Springer (2024)
15. Mokhtari, M., Ahmadi, N., Tsang, T.S., Abolmaesumi, P., Liao, R.: Gemtrans: a general, echocardiography-based, multi-level transformer framework for cardiovascular diagnosis. In: International Workshop on Machine Learning in Medical Imaging, pp. 1–10. Springer (2023)
16. Mokhtari, M., Tsang, T., Abolmaesumi, P., Liao, R.: Echognn: explainable ejection fraction estimation with graph neural networks. In: International Conference on Medical Image Computing and Computer-Assisted Intervention, pp. 360–369. Springer (2022)
17. Muhtaseb, R., Yaqub, M.: Echocotr: estimation of the left ventricular ejection fraction from spatiotemporal echocardiography. In: International Conference on Medical Image Computing and Computer-Assisted Intervention, pp. 370–379. Springer (2022)
18. Ouyang, D., et al.: Echonet-dynamic: a large new cardiac motion video data resource for medical machine learning. In: NeurIPS ML4H Workshop, pp. 1–11 (2019)
19. Ouyang, D., et al.: Video-based ai for beat-to-beat assessment of cardiac function. Nature **580**(7802), 252–256 (2020)
20. Reynaud, H., Vlontzos, A., Hou, B., Beqiri, A., Leeson, P., Kainz, B.: Ultrasound video transformers for cardiac ejection fraction estimation. In: Medical Image Computing and Computer Assisted Intervention–MICCAI 2021: 24th International Conference, Strasbourg, France, September 27–October 1, 2021, Proceedings, Part VI 24, pp. 495–505. Springer (2021)

21. Selvaraju, R.R., Cogswell, M., Das, A., Vedantam, R., Parikh, D., Batra, D.: Grad-cam: visual explanations from deep networks via gradient-based localization. In: Proceedings of the IEEE International Conference on Computer Vision, pp. 618–626 (2017)
22. Serrano, S., Smith, N.A.: Is attention interpretable? arXiv preprint arXiv:1906.03731 (2019)
23. Smilkov, D., Thorat, N., Kim, B., Viegas, F., Wattenberg, M.: Smoothgrad: removing noise by adding noise. arXiv preprint arXiv:1706.03825 (2017)
24. UBC Advanced Research Computing: UBC ARC sockeye (2019)
25. Vaseli, H., et al.: Protoasnet: dynamic prototypes for inherently interpretable and uncertainty-aware aortic stenosis classification in echocardiography. In: International Conference on Medical Image Computing and Computer-Assisted Intervention, pp. 368–378. Springer (2023)

© The Editor(s) (if applicable) and The Author(s), under exclusive license
to Springer Nature Switzerland AG 2026
M. Reyes et al. (Eds.): iMIMIC 2025, LNCS 16464, pp. 161–162, 2026.
https://doi.org/10.1007/978-3-032-17611-0

GPSR Compliance
The European Union's (EU) General Product Safety Regulation (GPSR) is a set
of rules that requires consumer products to be safe and our obligations to
ensure this.

If you have any concerns about our products, you can contact us on

ProductSafety@springernature.com

In case Publisher is established outside the EU, the EU authorized
representative is:

Springer Nature Customer Service Center GmbH
Europaplatz 3
69115 Heidelberg, Germany

www.ingramcontent.com/pod-product-compliance
Ingram Content Group UK Ltd.
Pitfield, Milton Keynes, MK11 3LW, UK
UKHW020819080726
473059UK00007B/2334